ECOLO
GARDENIN
FOR HOMI

Also by Lee Fryer & Dick Simmons

THE AMERICAN FARMER

EARTH FOODS

WHOLE FOODS FOR YOU

ECOLOGICAL GARDENING FOR HOME FOODS

Lee Fryer and Dick Simmons

NEW YORK 1975

This book is dedicated to MARTINE ANDERSON,
an ecological gardener who uses seaweed in growing gorgeous pest-free vegetables.

Printed in the United States of America
FIRST PRINTING

Library of Congress Cataloging in Publication Data

Fryer, Leland N 1908-
Ecological gardening for home foods.

Includes bibliographical references.
1. Vegetable gardening. 2. Organic gardening.
3. Vegetables--Preservation. I. Simmons, Dick, joint author. II. Title.
SB321.F88 635 74-34009
ISBN 0-88405-102-1
ISBN 0-88405-107-2 pbk.

CONTENTS

FOREWORD

Authorities and scientists have predicted food shortages in some countries and actual famine in others, which our present technologies will prove impossible to prevent. We have already seen the early stages of these shortages in 1974.

While we may not be able fully to prevent these catastrophes, we can alleviate their effects by producing part of our foods through personal and family gardens. A small piece of land 20 by 20 feet can partly feed two or three people or a small family.

Land is available almost everywhere we look. We find it in vacant lots, parks, and surrounding most cities as well as in country areas. We must learn to use it for home gardening—and youth gardening—as an important move in America to prevent local shortages of nutritious foods.

This book by Lee Fryer and Dick Simmons will probably be the definitive book to serve as a practical guide in this field, to answer questions and point the way toward lessening our food problems. With continually rising food prices, it shows a practical way to extend your food budget and to cope with inflation.

I urge that you read this book and pass it on. It is important for the physical health and economic welfare of every family in America.

EDDIE ALBERT

Pacific Palisades,
California

ACKNOWLEDGMENTS

Derek Fell, able Director of the National Garden Bureau, provided useful counseling as this book was produced.

Jeanne Davis, editor-publisher of Community Garden News, Columbia, Maryland, was especially helpful on community and neighborhood gardens. Her materials were supplemented by information and assistance from Gardens for All, a nonprofit group at Burlington, Vermont.

Bill Hash and Jerry Smith of Washington Youth Gardens, and Peter Wotowiec of the Cleveland Public Schools, helped with the youth gardening portions—with a big assist by Eddie Albert.

Robert Wearne and Russ Kaniuka of the U.S. Department of Agriculture provided quick access to good counselors and publications in their department.

Marian Fryer assisted in financing and counseling as the manuscript was written. Mildred Fryer provided inspiration and memories of excellent gardening and food storage. Tyree Nicholas supplied day and night stenography and a practical homemaker's advice. Sy Gresser coordinated research and graphics.

Nancy Davis, our editor at Mason/Charter Publishers, became a team worker with the authors and made springtime publication of this book possible.

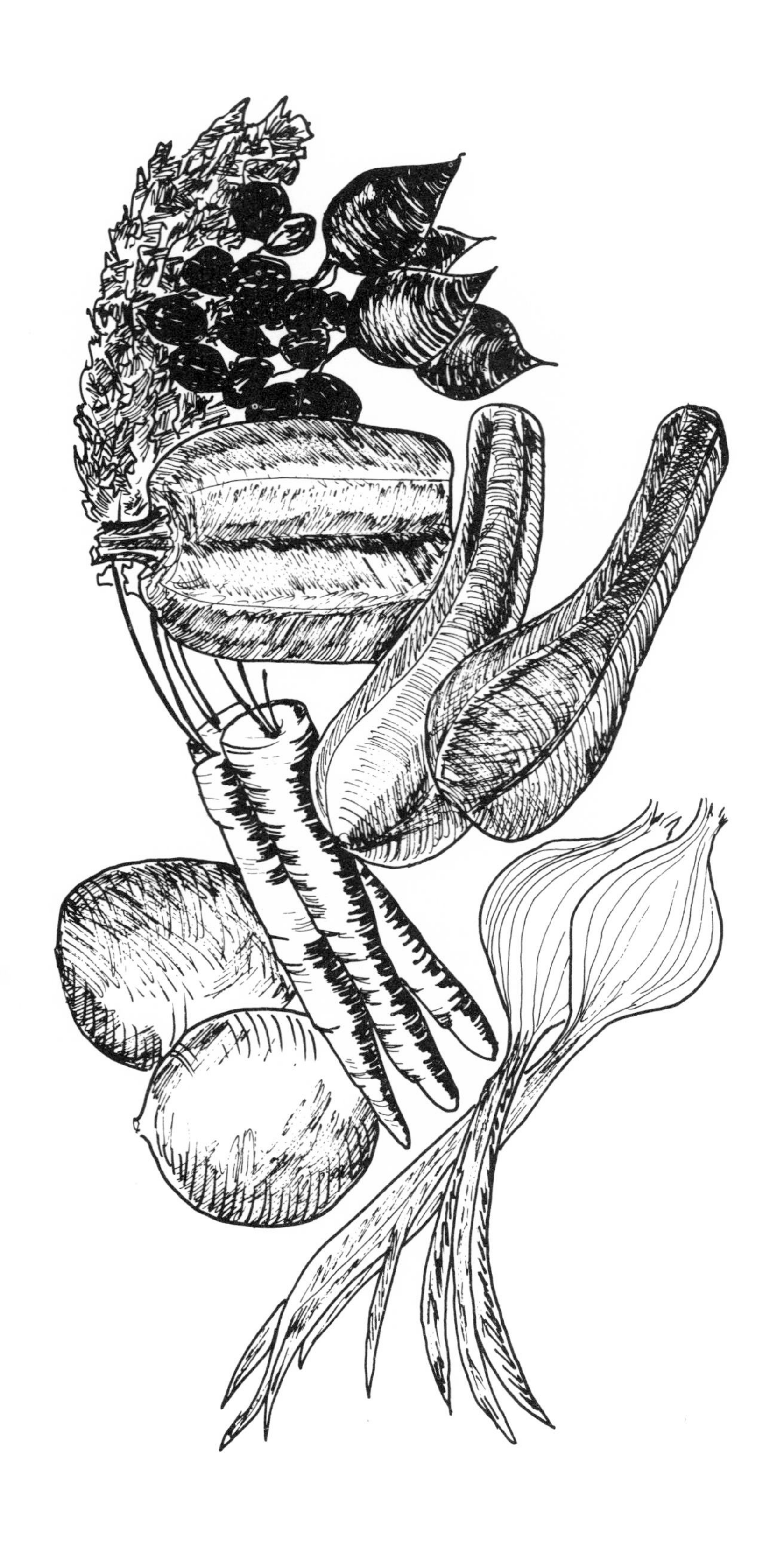

1.

FORTY MILLION GARDENERS CAN'T BE WRONG

Outside our window there is a tomato plant 8 feet high that is bearing twenty-eight young tomatoes. Since this is only mid-July, it will produce at least a dozen more—12 pounds of nutritious red-ripe fruit.

This tomato plant is growing in a flower bed beside the pansies, dahlias, and moss roses. There are four more tomato plants in other flower beds around the yard. Altogether they will yield over 50 pounds of superb tomatoes worth $20 at retail price in the supermarket. But they are better, much better, in flavor and quality.

You can do this same thing and more, in flower beds or in a larger vegetable garden, growing from $50 to $500 worth of tasty vegetables and fruits for direct home use. Doing it will not only help you beat inflation and high food costs, it will nourish your body and soothe your soul.

New gardens are blossoming all over America. Why?

- Because forty and fifty cents a pound seems too much to pay for common vegetables. True, people of many countries pay more than this, but Americans are resisting high food prices, and gardening is a constructive way to do this.
- Because home vegetables taste extra good as compared with commercial produce, which often cannot have full natural flavor, because it must be picked too soon and shipped too far.

- Because meat prices have risen nearly out of sight, and people wish to save on vegetables in order to buy regular quantities of meats.
- Because home and neighborhood gardening activities are satisfying and fun and good exercise.
- Because gasoline and traveling are too expensive, and people want more near-to-home recreation and personal projects.
- Because life in America is becoming too phoney and commercial, and gardening is a sensible, honest revolt against "the system."

Many people have found that gardening gives them a handle on the home food problem. From gardening they may go into making pickles, canning produce, buying meat at wholesale, and making their own bread. This is the new American trend, up and forward to a simpler life in which we do more things for ourselves and depend less on expensive gadgets and services.

You may be joining this sensible trend; if so, we hope the suggestions and guidelines of this book may help you.

WHY A HOME GARDEN?

In 1973 the Gallup Poll conducted a major study of home gardening in the United States for Garden Way Associates of Charlotte, Vermont. Knowing that about three million new gardeners entered the scene in 1973, Garden Way wanted to know what motivated people to grow their own foods, and how many seriously intended to do so.

Gallup's National Garden Survey found that economy—a desire to reduce food costs—was the main motivation for millions of people turning to gardening to grow more of their own foods. The survey also showed that about nineteen million additional Americans would grow gardens if they found they could save from $200 to $300 a year in food costs by doing so. The survey also found that getting good flavor and nutrition in home-grown vegetables was a strong additional motive for gardening. People love the true flavors of fresh-picked peas and sweet corn, of naturally ripened tomatoes and melons. They link flavor to good nutrition—if a vegetable or fruit tastes good, it must be good for you, providing more vitamins and minerals.

A Return to Nature

In addition, Gallup found that many Americans, particularly young adults between eighteen and twenty-nine, are turning to gardening as a way to "return to nature." They found a strong desire to escape from crowded conditions of cities. In fact, the findings of several surveys show that two-thirds of all Americans (68 percent) want more space. They would prefer to live in suburbs, small towns, or rural areas. As an intermediate step half of all Americans (54 percent) want to grow vegetable gardens.

The rise in home gardening, an increase of about 15 percent in 1974, is an expression of these deep-seated urges in American people. If you feel a strong desire to dig the soil, plant seeds, and "return to nature," you are joining a multitude of neighbors who have similar feelings.

"If Only I Had a Bit of Land . . ."

The National Garden Survey (Gallup) showed that about thirty million additional Americans (fifteen million households) would plant gardens if they had access to a bit of land. Of these, about eighteen million would be interested in neighborhood or community gardening if such a project became available to them, thus solving their land problems. A community garden is a place where a person with no land can travel a short distance and garden on a plot of soil. Such multiple-family gardens have become popular in Europe, and show signs of spreading into hundreds of American communities.

Derek Fell, Director of the National Garden Bureau, estimates that community gardening increased 50 percent in the United States in 1974. About thirty-five million people had individual home vegetable gardens. An additional million participated in community gardening, making a total of about thirty-six million Americans who grew some of their own foods in 1974.

How strong is this trend? How many people will be growning home foods by 1980? Our prediction is at least fifty million. The urge is present among a multitude of non-

Author Lee Fryer with one of the tomato plants grown in flower beds at his home. Seeds were sown in half-gallon cans in a sunny window in April and the young plant was transplanted into the fertilized flower bed at the end of May. This Big Boy tomato is one of several varieties you can grow with reasonable care and good feeding in your garden.

gardeners. Rising food prices and a desire for better living will stimulate a steady increase in home food production in the next five years.

FOUR KINDS OF FOOD GARDENS

Twenty years ago a garden was a garden, with neat rows of peas, corn, and cabbage—each garden was similar to the next one. Today people are growing lettuce in window boxes, string beans in tubs on rooftops, and tomatoes in flower beds. Small greenhouse units for growing vegetables are not unusual, and "salad gardens" are being featured in house and home magazines.

For purposes of the present book on home food production, we shall divide gardens into four kinds: minigardens, regular backyard gardens, community gardens, and hydroculture and greenhouse gardens.

Minigardens

There are many apartment dwellers and other urban people who do not have a plot of land but nevertheless plant seeds and grow a few vegetables. They use boxes or other containers to hold the soil, or they plant vegetables and salad gardens in their flower beds.

These minigardens are suitable for certain kinds of vegetables—lettuce, swiss chard, zucchini, radishes, green peppers, green onions, parsley, herbs, tomatoes, shallots, and even a few hills of cucumbers, potatoes, and strawberries.

In addition to providing personal pleasure and recreation, these minigardens may yield enough vegetables to save from $25 to $100 in yearly food costs.

Regular Backyard Gardens

There are many people who for years have gardened on plots of land spaded up or otherwise tilled in the spring. Such gardens may be as small as 10 by 20 feet or as large as half a city block.

The more spacious land area permits the growing of full assortments of vegetables and small fruits, including sweet corn, garden peas, beans, spinach, pumpkins, squash, cucumbers, melons, asparagus, cabbage, carrots, parsnips, turnips, collards, onions, potatoes, strawberries, gooseberries, currants, raspberries, and even dwarf apples and pears.

Such a garden may enable a family of four to save from $100 to $500 a year in food costs.

Community Gardens

There are more and more projects in which several people join together to obtain use of a piece of land for gardening. They usually subdivide it into individual plots, each person planting and caring for his own garden within the project. However, all or part of the group may plant corn, potatoes, or pumpkins together, sharing the harvest.

Included in this category are the gardening projects of urban young people, thousands of whom are successfully gardening on land provided by cities or civic organizations. For example, over a thousand boys and girls in Washington, D.C., were involved in the 1974 projects, each young person aiming to harvest at least 40 pounds of produce. We will describe such young people's gardens in Chapter 9.

Community garden plots may be large or small. They offer opportunities for substantial food cost savings plus a neighborly kind of recreation.

Hydroculture and Greenhouse Gardens

There are ultramodern kinds of vegetable gardens in which the gardener controls temperature, moisture, light, and nutrition of the crops in order to get high yields, good quality, and off-season supplies of vegetables.

In the case of hydroculture the plants are grown under a plastic roof in well-fertilized soil that is nourished with a light addition of plant foods to the irrigation water. High yields of top-quality vegetables are possible. For example, a 1,000-square-foot hydroculture garden may yield 5,000 pounds of tomatoes and cucumbers in a six-month season. Such a supply can serve a whole neighborhood or be the basis for a small family enterprise in growing and selling vegetables.

The small greenhouses are for people with good skills who wish to grow vegetables and flowers indoors. They may be as small as 20 or 30 square feet of growing area.

ECOLOGICAL GARDENING AND FOODS

For many years the favored mode of good gardening in America has been "organic." It is a system that makes use of "natural" fertilizers: compost, manure, bone meal, rock phosphate, pulverized rock, and other earthy materials. Poisonous weed sprays and pesticides are not acceptable in an organic gardening and food supply system.

Times have changed, however. Three-fourths of the United States population lives in towns and cities, where they have little access to manures and organic wastes and where it is very expensive and difficult to be a successful organic gardener.

As a practical alternative we suggest the materials and methods of *ecological gardening* and food production. It is an updated way to achieve the aims of organic gardening: to grow assuredly *safe* foods that are fully nutritious and to utilize and recycle organic wastes.

Ecological gardening has these good features:

- The fertilizers are balanced and mineralized, so they will grow healthy, full-flavored vegetable crops.
- Poisonous pesticides are not used. Pest control is achieved through the use of seaweed materials, biological controls, and good plant nutrition.
- The fertilizers and sprays are safe for pets, birds, and people. They will not pollute soil, water, or foods with harmful poisons.
- The gardening materials are medium-priced and available in most garden-supply markets.

Our work group has many years of experience in use of processed seaweed and fish materials in farming and gardening. In addition, through the courtesy of Clemson University and the University of Maryland, we have included their seaweed research findings in Chapter 5.

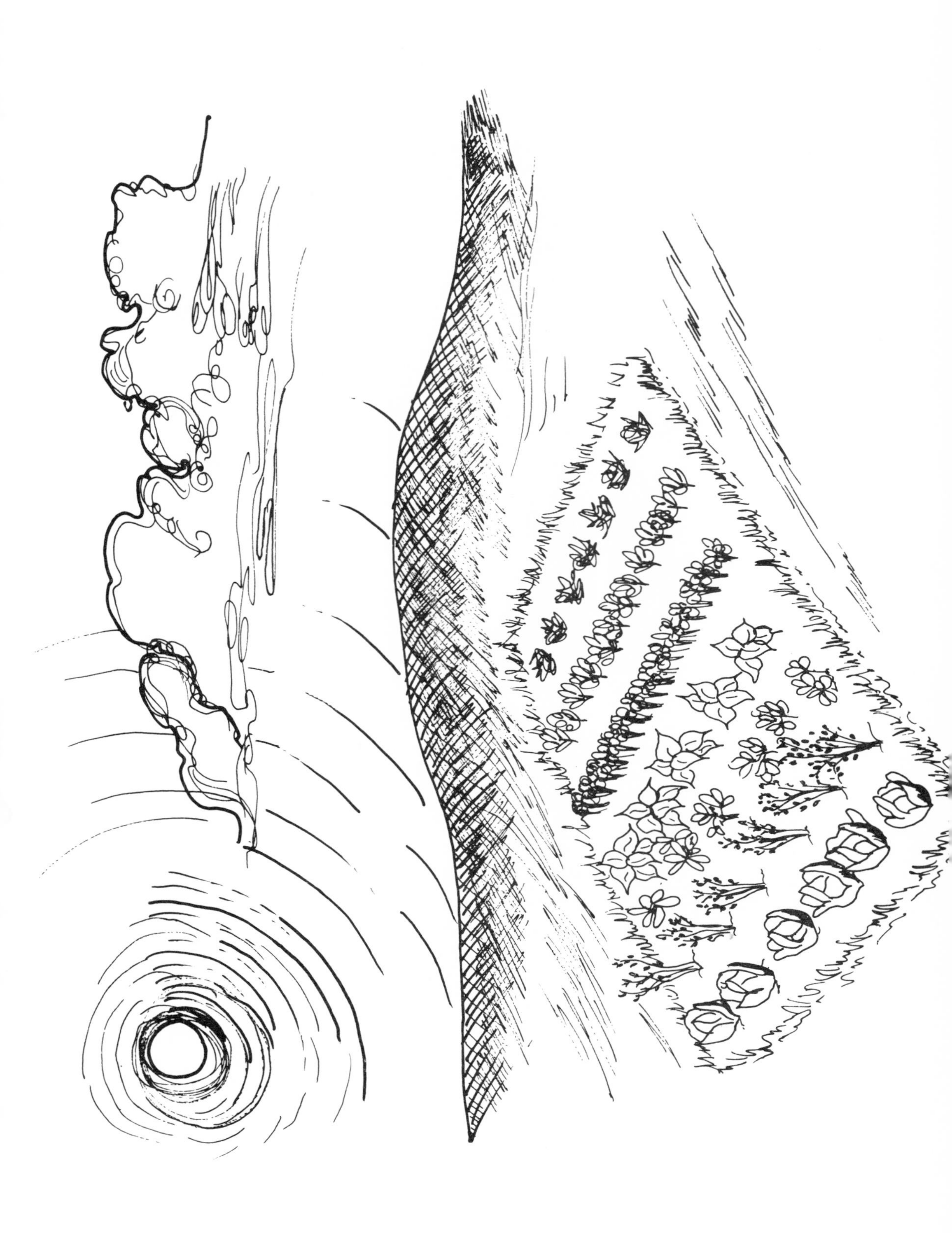

2.

A HOME GARDEN TO SAVE $200

Gallup's National Garden Survey showed that about half of all nongardeners in America—ninety million people—would like to grow gardens if this might save them at least $200 a year in food costs. Responding to this need, the National Garden Bureau has published a plan for a home garden to save $200 in food costs for a family of four. It was designed by Derek Fell, director of the bureau.

We have used this plan and Mr. Fell's suggestions in preparing the following design for a money-saving home garden. It shows what may be done by many home food producers to cope with inflation and the food budget crisis.

THE GARDEN AREA

To save $200 in food costs, you should have about 360 square feet of garden area; for example, a bit of land 12 by 30 feet. It is best to have it at least twice as long as wide and to plant the rows across the garden. In this way you can allot one or two 12-foot rows to each kind of vegetable. You may of course adjust the garden space to your property. If a square garden is better, do it that way—make it 20 by 20 with a path across the middle. Both of the gardens illustrated on the following page have 360 square feet of gardening space.

The first garden has rows 12 feet long, a good length for many kinds of vegetables. With rows 2 feet apart and some succession planting (such as corn following lettuce), you can grow eighteen or twenty different kinds of vege-

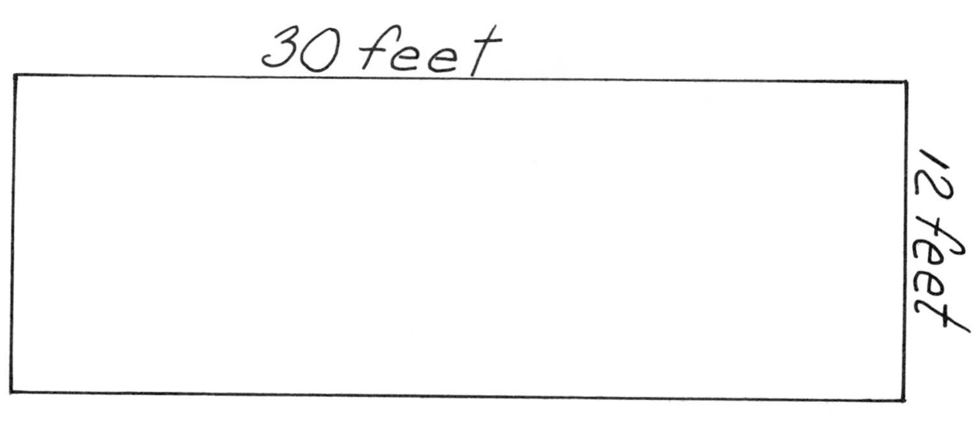

Long and narrow garden

tables. Also, you can walk around your garden and take care of it with a minimum amount of tramping on the soil on wet days.

The second garden has a path 2 feet wide down the middle, and rows 9 feet long. Two or three rows can be

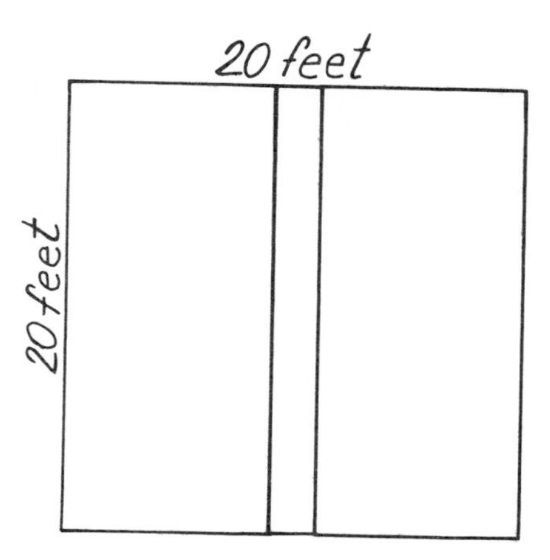

Square garden

used for a main vegetable—for example, peas or beans—and one row for a minor crop, such as parsley or green onions. The path down the middle enables you to tend the garden and show it to the neighbors with a minimum of difficulty.

If in Doubt, Make It Smaller

Inexperienced gardeners often make the mistake of tilling and planting too much area, failing to do a good job in preparing the soil and tending the plants. Or they try to squeeze a 300-square-foot garden into a 400-foot backyard, when half the size would be better. Or they plant a big garden in a shady area where sun-loving plants could not possibly do well. Or they plant a big garden on rough soil—which has not been cultivated before—and half the garden dies due to poor soil.

A smaller garden the first year may be a much better project, giving you a taste of success and useful experience for use in future gardening years.

So . . . if in doubt, make your garden smaller. Save $100 in grocery costs instead of trying for $200 and saving only a nickle or a dime.

LOCATION

The next step in planning your garden area is to consider the various conditions that may influence your success or failure.

Sun

Vegetable plants need sunshine as their power for food-making and growth. Such leafy kinds as lettuce and chard may do fairly well in medium shade, but as a general rule, plant your garden where it will have sunshine at least half the day.

Trees

Nearby trees may cause too much shade; also, their roots may extend under your soil area, competing with your garden for plant foods and especially for moisture. If so, your garden may grow well for a month or two, then get puny because the trees are taking the moisture and some foods. Trees with shallow roots are big competitors. Try to avoid this problem by placing your garden away from trees, but if this is difficult, prepare to fertilize and irrigate enough to compensate for the tree roots.

Grass

If your garden area was cultivated last year, the soil may be in good condition, fairly free of grass and easy to prepare for this year's planting. However, if the area you've selected

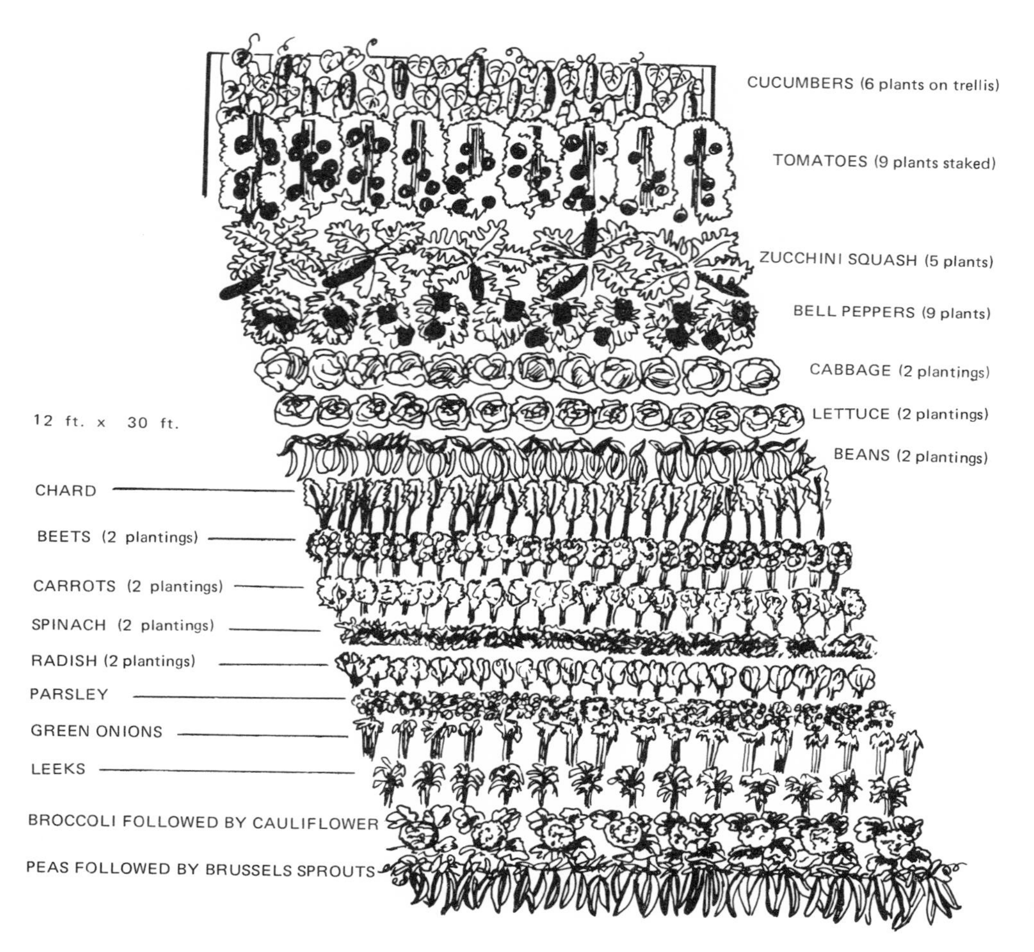

Model garden plan based on design by National Garden Bureau

for your garden is a grassy sod, you'll have to be a pioneer —a "sod-buster." Take this into account in choosing your garden area and making plans. Later, under Soil Preparation, we will tell you how to be a sod-buster, if that becomes necessary.

THE GARDEN PLAN

Now here we go: fifteen or twenty kinds of vegetables, all in rows. What shall we plant? How much of each kind? When?

The National Garden Bureau is in touch with gardeners and seed companies all over the country. Thus, their suggestions are useful, and we will use the Garden Bureau's selection of vegetables as a flexible base—to be adapted to your area and your personal tastes and needs.

The model garden plan for 300 square feet of ground includes twenty-one kinds of vegetables. You may grow all of these varieties, or you may choose more of the basic ones and fewer or none of the others. For example, if you love juicy garden-fresh corn, you may have three rows of it, cutting out the cauliflower; you may increase the early peas and omit the early cabbage; you may even add a row of ever-bearing strawberries or eggplant.

Remember: If you are inexperienced—an Inflation Age pioneer—you may be well advised to plant a smaller garden, with only half as many kinds of vegetables as those in this model garden.

SOIL PREPARATION

Good spring gardens are often prepared the previous fall by a thorough spading or tilling of the soil. This may accomplish two objectives favorable to success: breaking up grassy sod and other debris, mixing it with soil so that during the winter it may decompose into usable organic matter; breaking up and aerating hard (compact) soil, so that it may be more easily tilled and planted the following spring.

Fall tillage also provides an opportunity to add compost, leaves, manure, fish, seaweed, peat moss, and other organic materials to the garden bed to rot and percolate all winter, adding fertility to the garden. This is your secret weapon to

DOUBLE-CROPPING INCREASES THE VALUE OF YOUR GARDEN

To gain maximum value from a vegetable garden "double-cropping" is essential. By "double-cropping" we mean gaining two plantings to each garden row—one to mature in summer, the other in fall.

For example, spinach is an early crop—it requires sowing as early in the spring as possible, so it matures before hot weather sets in. This will leave a space for another planting later in the summer to mature in fall when cooler weather returns.

Peas are another crop that bear early and are finished by early summer. That same space can then be dug over and replanted with another row of broccoli, cauliflower, or Brussels sprouts to mature during fall of the same year.

Corn reaches a peak in mid-summer, and the space it occupies can be replanted with a fall crop of cabbage, lettuce, or broccoli.

The following are some varieties of vegetables which can be double-cropped by a planting in spring followed by another in mid-summer for fall harvesting:

Lettuce, beets, carrots, spinach, turnips, bush beans, broccoli, cauliflower, cabbage, and radish.

Certain vegetable varieties do better as a fall crop than a spring or summer crop. These include broccoli, cauliflower, and Brussels sprouts. They prefer to be planted in mid-summer so they mature during the cool weather of fall. Brussels sprouts planted this way are especially valuable, since the green sprouts can be picked into December over most of the United States when store prices for green vegetables are high.

Other vegetables take a long growing season to mature properly. These include leeks, pumpkins, winter squash, salsify, and parsnips, so double-cropping with these is not possible.

The best group of vegetables to grow in a home vegetable garden are those which mature reasonably early and stay productive over the entire gardening season, surviving a wide range of conditions. These superproductive vegetables are zucchini squash, chard, parsley, tomatoes, and peppers.

Derek Fell, Director
National Garden Bureau

be used in growing exceptionally fine vegetables: start your action the previous fall.

However, if fall work on your garden is impractical or impossible, do not feel frustrated. Do the next best thing and spade or till your garden area quite early in the spring. Again, add compost, manure, peat moss, or other organic materials to promote fertility. When added early, they decompose and help to nourish your vegetable crops.

How to Spade or Till the Soil

For those beginners who do not know the language, spading the soil is turning over the top 8 inches of soil—in effect, "plowing" it in preparation for planting. This operation may be done with a No. 2 shovel, a lighter shovel, or a spade. Ask your hardware or garden store personnel for advice about shovels, spades, and spading.

Take shovel or spade in hand and press it into the soil with your foot, then lift and turn over the chunk of soil. Then you *whack it* with the back of your implement to break and pulverize this bit of earth. A well-spaded garden looks light and loamy. If the chunks are wet and sticky, you are spading it too early—or too soon after a rainy spell. The soil should crumble when hit with a shovel.

If you prefer, you may rent (or buy) a rototiller for use in "plowing" your garden area. They are available at most hardware and garden stores and machinery rental places.

Under average conditions it will take you about four hours to spade a 360-square-foot garden area—that's half a day of pleasant work and exercise. Oh, yes, you can brag about spading over beer or the back fence: An acre of loam soil 6 inches deep weighs two million pounds. Your 360-square-foot garden is about 1/120 of an acre. Therefore, its top 6 inches weighs about 8 tons. That is what you lift and turn over, if you spade your own garden.

Sod Busting

If your garden area is in grassy sod, you have a special problem getting it in shape for planting. Besides "plowing" it by spading or tilling, you must also successfully rot the sod so it becomes part of the soil. Unless you do this, two adverse things may happen: the unrotted grass and roots will make the soil too fluffy, causing your garden to dry out; the rotting of the sod during garden season will demand nitro-

ROTOTILLERS FOR HOME GARDENING

Plowing the garden was the early way to turn the soil. The sharp blade of the plow lifted a piece of soil 8 inches deep and a foot wide and turned it over, with grass, trash, roots, and debris on the bottom of the furrow.

Rototilling is the up-to-date way. It uses a small engine and rotating blades to plow, mix, turn, and compost everything in the top 8 inches of soil—or deeper if desired.

A good garden rototiller has an engine, two wheels, a rotating tiller with blades to cut the soil, two handles, and control levers. Since the wheels are powered to move the machine ahead, it is easy to operate it without getting very tired.

The rotating blades should be *behind* the wheels, at the back of the tiller, in order to do a good job without tiring the operator. This puts the weight of the machine fully into the process of tilling and mixing the soil. Several different brands are designed this way.

Rototilling takes the place of plowing or spading the soil, and it does a splendid job. In a typical operation it mixes top growth, roots, leaves, and fertilizer with the soil, making a compost within the garden. You can spread extra leaves, trash, stalks, garbage, seaweed, fish, bones, lime, gypsum, and even feathers on the garden and rototill it all in.

These useful implements cost from $500 to $700 and weigh from 300 to 500 pounds. If well cared for, they will give good service for fifteen years. Rototillers are great for people with fairly large gardens (over 500 square feet) and for groups of people with land for a neighborhood garden.

gen, tending to deprive your vegetable plants of this necessary food.

This is another good reason to spade or till the grassy area in fall, if possible, or at least in early spring. Then, before turning (or tilling) the soil, add 3 pounds of mixed fertilizer per 100 square feet of area to help in rotting the sod—that's about 10 pounds for a 360-square-foot garden.

If the sod is quite heavy, it is advantageous to use a rototiller. The machine will break the sod and mix it with the soil and fertilizer better than spading. However, spading is cheaper.

After "sod busting," spade or till the soil again before planting. Otherwise it may still be too rough for good gardening.

Lime

Soil tests are useful to determine your needs for lime and fertilizer. However, getting the tests may be difficult and time consuming. In the absence of tests you may use this guide: if the rainfall in your area is more than 25 inches a year, add lime to your soil at the rate of 10 pounds per 100 square feet—about 30 or 40 pounds for a garden of 360 square feet. Apply it before spading or tilling. For lime requirements in other areas see the section on liming your soil in Chapter 4.

Use agricultural lime or dolomite (lime with magnesium in it). If you cannot afford lime, use ashes or omit the lime. While lime helps to grow a better garden, it is not indispensable.

Organic Matter

Adding organic matter to your garden area will promote success. Use any kind available—compost, manure, leaf mold, straw, seaweed, leaves, fish wastes, feathers, tankage, or peat moss. As a general guide use 10 to 20 pounds per 100 square feet of area—about 50 pounds for a 360-square-foot garden.

Add the organic materials at three different times:

1. Before spading or tilling, so as to mix with the soil.
2. Before planting, by cultivating into the soil.
3. As side-dressing alongside plants and rows.

As noted before, do not apply unrotted organic materials during the growing season. The rotting process may demand nitrogen, depriving your vegetable plants of their normal food. Compost the organic materials ahead of time, if possible, or purchase dry manure, compost, or garden mulch.

If you cannot afford to buy organic materials, try to gather some; if that is difficult, you can do without. People have been growing gardens for centuries making do with whatever they could get their hands on.

Fertilizers

This subject is covered in Chapter 5, How to Make and Use Good Fertilizers. You may obtain basic knowledge from that section, meantime using the following guides:

When Spading or Tilling. In the spring add 4 pounds of mixed fertilizer per 100 square feet of garden area, broadcast on the soil before tillage. This is about 15 pounds for a 360-foot garden.

When Planting. For rows of seeds or transplants dig the furrow 1 inch deeper than usual. Sow a light sprinkling of mixed fertilizer in each furrow, so it will be under the seed. Then cover the fertilizer with soil, restoring the furrow to normal depth, and sow seeds as usual. This added food under the rows will help to make a successful garden. But go easy; a scant cup of fertilizer per 12-foot row is sufficient.

When planting hills of seeds, such as zucchini or cucumbers, add a handful of fertilizer within the hill, mixed into the soil. Use the same procedure when transplanting tomatoes, peppers, and other garden plants.

Side-Dressing and Liquid Fertilizer. As your plants grow, you may wish to encourage high yield and quality by added feeding. If so, you may side-dress the rows and plants with an occasional light application of mixed fertilizer, or you may make gentle applications of a diluted liquid fertilizer by spraying it on the leaves of the plants.

Many skilled gardeners use these methods of supplemental feeding.

PEST CONTROL

The guides and suggestions for protecting your garden from insects and plant diseases are given in Chapter 6. As

GOOD TIMING IN GARDENING OPERATIONS

Gardening is seasonal, linked to supplies of sunlight, moisture, warmth, and the habits of plants in responding to changing weather. In our climate zones the best time for land preparation is the fall; the best time for seed selection and planning is winter; and the main time for planting is spring. Of course, spring begins in February in many states with mild climate and begins in May in Alaska. Good gardeners learn their own climates and read the weather guides.

Some of the best gardens are fertilized and tilled in October, with the addition of rough organic wastes that will rot during winter and early spring. If you moved to your place on January 1st, however, do not be dismayed. Land preparation early in spring will usually suffice.

After taking into account climatic differences, we should remember that different kinds of plants have their own relations to sunlight. This is called *photoperiodism*. It means that a given kind of plant—garden peas, for example—knows when to wake up, grow, bloom, bear seeds, and go to sleep again. In the case of peas it is early. There is not much use in planting peas after mid-spring; they will not do well. Spinach is another early-season crop. On the other hand, corn and potatoes may be planted at various seasons of the year, if you select suitable varieties. The authors have had fresh garden corn from mid-July until late September by choosing different varieties and making succession plantings.

For further information on this phase, feel free to write Derek Fell, Director, National Garden Bureau, Gardenville, Pennsylvania 18926. Or, visit your county's Agricultural Extension Office.

you will discover, strong healthy plants that are fed with balanced fertilizers (not too much nitrogen) will resist attacks by pests, while puny and underfed plants are eaten by every bug in the neighborhood.

In ecological gardening we do not try to poison and exterminate all weeds and insects, but rather to use methods and materials that enable vegetable plants to outgrow and resist pests. This strategy is generally more successful than "spray and kill" methods, and it surely is safer.

BUDGET AND FOOD COST SAVINGS

When well planted and tended, the model home garden has the power to save a family of four persons at least $200 in food costs. Here is the financial analysis:

Costs			
Seeds, 20 kinds at an average cost of 50¢ per packet or quantity		$10.00	
Production materials			
Lime, 25 lbs.	$1.00		
Mixed fertilizer, 25 lbs.	4.00		
Organic materials	3.00		
Pest control	2.00		
Total		$10.00	
Tools and equipment			
This year's share of $15.00 cost		$5.00	
Miscellaneous		$5.00	
Total Costs			$30.00

YIELDS & VEGETABLE VALUES*

60 pcs. cucumbers @ .20 each	$12.00
90 lbs. tomatoes @ .45 lb.	40.50
30 lbs. zucchini @ .30 lb.	9.00
30 lbs. peppers @ .40 lb.	12.00

*Based on estimates of National Garden Bureau.

24 hds. cabbage @ .60 head	14.40
40 hds. lettuce @ .40 hd.	16.00
30 lbs. string beans @ .40 lb.	12.00
30 lbs. chard or collards @ .45 lb.	13.50
30 lbs. beets @ .30 lb.	9.00
40 lbs. carrots @ .30 lb.	12.00
12 lbs. spinach @ .45 lb.	5.40
24 bunches radish @ .30 bunch	7.20
20 bunches parsley @ .30 bunch	6.00
24 bunches green onions @ .20 bunch	4.80
20 bunches leeks or celery @ .45 bunch	9.00
25 lbs. broccoli @ .40 lb.	10.00
12 hds. cauliflower @ .60 hd.	7.20
15 lbs. peas @ .40 lb.	6.00
30 pts. brussels sprouts @ .45 pt.	13.50
40 lbs. potatoes @ .15 lb.	6.00
30 ears corn @ .15 ear	4.50
Total value of crop	$230.00
Deduct costs	30.00
Net savings in food costs	$200.00

HOW TO USE VEGETABLES FROM YOUR GARDEN

During an earlier time in America over half of the people lived in small towns and rural areas. In those days almost everyone knew how to store and use foods to get the most out of what was available. It was not uncommon for a family to can 400 quarts of fruits and vegetables and make 5 gallons of pickles and sauerkraut. A homemaker would see a head of cabbage in four dimensions—to eat fresh in salad or coleslaw, to boil for dinner, to store in the cellar for winter eating, and to use in making sauerkraut.

Today, as food prices rise still higher, we need to regain some of these skills in order to extend the useful life of garden produce—and of meats, fish, and fruits that may be obtained in quantity or at wholesale costs.

The yields of vegetables from our model garden lend themselves to multiple uses. You will notice that there are five different ways these vegetables may be used. Canning, freezing, and storing can increase—even double—the value of your home vegetables. And the same skills and equipment

USE OF GARDEN VEGETABLES

VEGETABLE	SALADS	COOK-ING	CAN-NING	FREEZ-ING	STOR-ING	REMARKS
Cucumbers	yes		yes	—	—	Especially good as pickles
Tomatoes	yes	yes	yes	yes	—	Green tomatoes picked before frost will ripen indoors during winter
Zucchini	yes	yes	yes	yes	—	Can be used as substitute for cucumbers in salads
Peppers	yes	yes	yes	yes	—	Especially good stuffed with meat
Cabbage	yes	yes	yes	yes	—	Makes good sauerkraut for freezing
Beans	yes	yes	yes	yes	yes	Dried beans store well for winter use
Chard	yes	yes	—	yes	—	Very hardy; lasts into winter months
Beets	if cooked	yes	yes	yes	yes	Will store through winter in a box or moist sand in cool basement
Carrots	yes	yes	—	yes	yes	Will store through winter in a box of moist sand in cool basement
Spinach	yes	yes	—	yes	—	Grows quickly during cool weather of spring and fall

Radish	yes	—	—	—	—	Can be braised to make a cooked vegetable
Parsley	yes	—	—	yes	—	Used mostly as garnish
Green Onion	yes	—	yes	yes	—	Dried onions will keep during winter in a dry cool place
Leeks	—	yes	—	—	yes	Will keep during winter in a box of moist sand in cool basement
Broccoli	—	yes	yes	yes	—	Plants grow one main head, and side shoots grow smaller heads
Cauli-flower	yes	yes	yes	yes	—	Best grown as a fall crop
Peas	—	yes	yes	yes	—	Edible podded peas also good to grow
Brussels sprouts	—	yes	yes	yes	—	Best grown as a fall crop; lasts well into winter
Corn	—	yes	yes	yes	—	May be sun-dried for winter
Potatoes	—	yes	—	—	yes	Will keep if stored in a cool, dry place

may serve you in curing, freezing, canning, and storing meats, fish, poultry, and fruits.

Be a pack rat and a hoarder. Gather foods and store them for winter eating. Doing this may keep you fat and philosophical even in this inflationary age.

PRACTICAL CHOICES

Admittedly, this model home garden to save $200 in food costs is a major family enterprise. Growing it successfully takes skill and lots of work—about 120 hours (fifteen days) of family labor in planning, spading, preparing, planting, tending, replanting, and harvesting work. Then there's more work in processing and storing the produce.

Many readers will prefer to expend less effort and even less money. Also, some readers have never gardened before, and they may wish to learn with a smaller, simpler project in the first year.

We suggest the following choices for consideration by such people who want a less ambitious first-year garden.

1. *Reduce the Number of Kinds of Vegetables.* Instead of twenty, plant only eight or ten kinds of vegetables—for example, peas, beans, potatoes, corn, lettuce, radishes, cucumbers, zucchini, and tomatoes. These are old familiar kinds, easy to grow and readily used by family and friends. Learning with these will qualify you for a more complex garden next year.

2. *Reduce the Size of the Garden.* Prepare and plant a 200-square-foot garden, and expand it to 300 or 400 feet next year.

3. *Reduce the Rate of Fertilizing.* The rate provided is a commercial gardener's, used to achieve high yields. It can be cut in half and still give good results. However, this saves only $5. We recommend this choice only if you "don't know what in the world to do with all those vegetables."

4. *Reduce Costs by Joining with a Neighbor.* You may be able to cut the cash costs in half by joining with a neighbor or friend. Seed packets often contain enough seeds for two families. Also, the tools and equipment may be shared.

This may be a way to finance the garden on a tight family budget.

In later chapters we will discuss minigardens and various kinds of community gardens.

3.

HOW PLANTS GROW

Three hundred years ago a Dutch botanist named Von Helmont pondered the question: if I grow a plant and it weighs 100 pounds, where does this mass come from? Does it all come from the soil?

To find an answer Von Helmont performed an historic experiment. He baked soil in an oven to remove moisture, then weighed out exactly 200 pounds of it into a big earthen pot. In it he planted a small willow tree trimmed to weigh exactly 5 pounds. Von Helmont then grew the tree for 5 years, providing no manure or fertilizer—only water at regular intervals. He then weighed the tree and the soil in which it grew. The tree had gained 164 pounds but the soil—dried again and carefully weighed—had lost only 2 ounces.

Where did the 164 pounds come from? Von Helmont said:*

> In the end I dried the soil once more and got the same 200 pounds that I started with, less about 2 ounces. Therefore, the 164 pounds of wood, bark and roots arose from the water alone.

Of course, Von Helmont was wrong. The gain in weight came mainly from the air via the process of photosynthesis, not from water. But this inquisitive man stimulated labora-

*From *Ortus Medicinae*. Amsterdam 1652. As reported in *Soil Conditions & Plant Growth* by Sir John Russell. (London: Longmans, Green & Co., 1950.)

tory research in this field and proved that plants derive only a small part of their substance—the minerals—from the soil.

If you bake garden plants to remove the moisture, about 90 percent of the remaining dry matter comes from the air and only 10 percent from the soil. The leaves capture the 90 percent portion of the dry weight from the air by means of photosynthesis.

PLANTS ARE ENERGY MAKERS

In our energy-troubled times it is useful to know that garden plants seize energy from the sun and store it in the form of sugar. Then they release and use this new energy in building all kinds of cells and tissues, in producing growth. The sugar-forming process is called photosynthesis; the energy-using one is called respiration. Chemists express it this way:

In photosynthesis:
energy + water (H_2O) + carbon dioxide (CO_2) =
sugar ($C_6H_{12}O_6$)

In respiration:
sugar ($C_6H_{12}O_6$) + oxygen (O) =
carbon dioxide (CO_2) + energy

Thus, plants receive infrared light waves in the chlorophyll cells of their leaves—that's the green matter in them. Using this energy along with air and water, they make sugar by reassembling carbon, hydrogen, and oxygen in a different chemical structure. It is the primal act in forming and sustaining life.

Sugar is to plant life as gasoline is to automobiles—the fuel and energy source. With sugar, plants can form new cells and grow. However, it is noteworthy that earth minerals are also essential in the growth of plants. Even the chlorophyll cells of leaves cannot be formed without atoms of magnesium from the earth.

MINERALS FROM THE SOIL

Typical plant cells contain carbohydrates—the CHO units derived from sugar—plus different combinations of earth minerals added. Calcium and boron are needed for cell walls. Phosphorus is used to make the nuclei of cells and to form seeds, flowers, and root tips. Potassium is a mobile element always present in sap and essential in forming strong tissues of all kinds. Iron serves as an oxygen carrier in plants, just as it does in life systems of birds, animals, and people.

Nitrogen, which plants use copiously in their growth, is an element of the air; however, plants are unable to take it directly from the atmosphere. Soil bacteria assist in "fixing" nitrogen so that plants can draw it in through their roots in acceptable forms. Manures and decaying vegetation —and commercial fertilizers—add to these soil sources of nitrogen.

As many readers know, plants require relatively large amounts of nitrogen, phosphorus, and potassium for their growth. These elements are shown on fertilizer packages in numbers such as 5–10–10 and 6–10–4, which signify the percentages of nitrogen, phosphorus and potassium, in that order, in the fertilizer products.

Plants also utilize whole assortments of earth minerals in addition to those named above—traces of minor elements such as zinc, copper, manganese, molybdenum, and selenium. There are at least fifty active minerals in our environment, and all of them are found in the tissues of various kinds of plants. Seaweed contains all the known minerals, while land plants contain twenty or thirty of them.

We must assume that most of these minerals, along with sugar, have roles of one kind or another in forming cells and tissues of plants.

ACTUAL CELL BUILDING USING GENE STRIP PATTERNS

The various parts of a plant—leaves, roots, flowers, scent, seeds, and all the rest—are amazingly true to type for any individual example of a species. For example, grass leaves always have parallel veins, while rose leaves have branched or netted ones; peaches always have only one seed rather than a dozen or more seeds per fruit, like apples.

How can all this happen? How do roses, grass, apples, and peaches know exactly how to form their cells and grow according to their family specifications? If you are a dressmaker or a computer operator, you will immediately understand the answer—patterns.

Every plant and creature has a pattern, an inherited gene strip that determines the size, shape, color, functions, and other features of its cell structures. In the case of plants this gene strip is transmitted through seeds to the new generation of the species, as a pattern for making and shaping the cells of the new individual.

The plant has *enzymes*, little protein units, which act as engineers on the scene of cell-building action. They receive dozens of raw materials, including sugar, and put them in place as new plant structures, building cells in sap, leaves, stems, roots, buds, flowers, fruit, and new seeds to perpetuate the variety or species. The story of enzymes is, in itself, a miracle tale—how they do their work by creating "hot spots" to attract essential cell-building materials, such as nitrogen and iron, and then linking the new molecules together in new plant cells.

Enzymes superintend their own construction, forming one another. There are thousands of different kinds of them, each requiring a tiny particle of an earth mineral as the core of its chemical structure.

Within this beautifully complex yet well-organized system, is it any wonder that soil and plant scientists are still uncertain how many different minerals are needed in normal plant growth, and exactly how they are all utilized? Artists are reluctant to gild lilies; scientists are modest in telling how to change the scent of a rose or the flavor of a sliced tomato.

TALL TALES ABOUT GROWING PLANTS

At least eighty different foods, factors, and conditions influence how plants grow. The following tall tales illustrate some of these. We could recount a dozen more.

Cuzco Corn

In 1959 at Seattle, Washington, the authors received four kernels of Cuzco corn seed from Peru, of a variety grown centuries ago by Inca Indians living at an altitude of about 10,000 feet. The kernels were enormous, about the size of an elk's tooth.

In a plot of fertile garden soil we planted each seed lovingly and added a cup of fertilizer similar to the *De-*

luxe Eco-Grow described in Chapter 5. Three of the seeds sprouted, but the fourth was sterile and never made a plant.

As this corn exploded from the soil, we knew a miracle might happen. Even as seedlings the plants were giants. Soon the leaves were 8 inches across and the stalks began to look like small trees.

In our mild Seattle climate at sea level, the Cuzco corn grew to a height of 18 feet. It produced brace roots like those of cypress trees from each lower joint to the ground, to avoid toppling over in the wind. The stalks were 4 inches in diameter by August first.

The Cuzco corn grew ears away up there 9 feet above the ground. We could barely reach the bottom one with a hoe. In our climate the ears never ripened so we could have more seed. Full of Inca dreams and memories, this Cuzco corn just kept on growing until time of frost.

How would it have grown on the *altiplano* of Peru? We do not know; plants perform differently in various climate zones.

Arctic Corn, Beans, and Potatoes

During 1950–1960 we studied arctic vegetables while supplying Alaskans with garden fertilizers. Many kinds were grown at the University of Alaska Experiment Station farm near Fairbanks. One variety of corn imported from Siberia grew flat on the earth, making tassels and ripe ears in that ridiculous position, using every erg of warmth the soil could provide. This corn was smart enough to avoid cold arctic breezes.

Arctic string beans grew 8 inches high and bore 6-inch pods abundantly. The plants knew they had little time between frosts to reproduce, so they wasted little energy making stems and leaves, concentrating on quick seed production. The quality of the beans was superb. During the long arctic summer days the plants make sugar like mad and put it into the vegetables.

Dr. Curtis Dearborn, Alaska's cold-weather plant breeder, developed potatoes to withstand 25° Fahrenheit—that's 7 degrees of frost. His finest achievement was a hybrid potato variety to yield big crops in the Alaskan climate. Many farmers raised these Dearborn potatoes, which grew 30 inches high and had lovely lavender flowers. These plants of great vigor and beauty each yielded six or eight smooth potatoes. Seen in the midnight arctic sun, such a field soothes the soul and lives forever in memory.

We carried Dearborn potato seed back to Seattle's gentle climate and grew it in the garden. What a difference! The plants looked like juvenile delinquents—recessive, inadequate, and hardly grew a tuber.

Sun, heat, length of day, gravity, and the earth's rotation all influence how plants grow.

Cucumbers Waiting for Magnesium

One day we had a desperate call from an Italian market gardener near Puyallup, Washington. He was in danger of losing his crop of greenhouse cucumbers and a large income because of a strange malady. The cucumbers had grown variegated leaves, yellow and green, like coleus plants.

To a plant scientist this says something, since the classical symptom of magnesium hunger is a loss of green color in the leaves, between the veins. Without magnesium the green chlorophyll cells cannot be formed, the leaves get yellow blotches, and growth of the plants ceases.

A visit to the greenhouse showed this was probably an accurate diagnosis. Therefore, we had this market gardener buy 50 pounds of epsom salts, which is magnesium sulfate ($MgSO_4$). He dissolved this in water and sprayed his crop. Within a week the green color was back; gorgeous new leaves and cukes were growing again.

While waiting for magnesium, these plants could not grow; for without magnesium they could not make chlorophyll, and without chlorophyll they could not make sugar.

NUTRITIONAL RELATIONSHIPS

Plants, and creatures that eat plants, are the main sources of human foods. If the plants contained only nitrogen, phosphorus, and potassium, we would be poorly nourished; in fact, we would dwindle and die of malnutrition. Even if we received all the minerals in plants, we would eventually die, due to a lack of cobalt and iodine, which plants do not require for normal growth but which people need.

The truth is gradually penetrating nutrition science that

people need more minerals in their foods than farmers need for high yields of good-looking crops. We may need full assortments of earth minerals in our foods in order to resist illness.

Thus, gardening should be seen as an opportunity to grow fully mineralized foods that contribute to personal health and well-being—particularly at the present time, when much of the supermarket produce is anemic, unripe, devitalized, and poorly grown.

One reason for learning how plants grow is in order to grow them better for human consumption. We need full-flavored, well-mineralized "health foods" in the best meaning of that term.

MOVEMENT OF NUTRIENTS IN PLANTS

In understanding how plants grow, it is also necessary to visualize the movements of sugar, minerals, water, and other substances within the plant.

First of all, the newly formed sugar in the leaves must move to sites of cell building throughout the plant, even down into the roots. This is accomplished through special phloem vessels, which might be compared with arteries in the human body, carrying oxygen-laden blood to all parts and tissues. In the case of plants the phloem vessels extend from the leaves to the smallest roots. In a tree 80 feet high these vessels are 80 feet long *plus* the distance to the ends of the roots. The phloem vessels handle downward movement of sap along with the distribution of energy-charged sugar to all parts of the plant.

The xylem tissues, on the other hand, handle upward movement of water, soil minerals, and nitrogen. These substances are absorbed through the roots, and the minerals and nitrogen are dissolved in the water as constituents of upward-moving sap.

Transpiration, which is evaporation of water from the plant's leaves, creates the suction force to move the sap upward, carrying food-laden water to all active areas of cell building, of plant growth. Here, under supervision of the enzymes, sugar-laden sap from leaves meets mineral-laden sap from roots, and blessed events occur. New cells are formed, and the plants grow.

EPILOG: PLANTS, YOU, AND HISTORY

The life processes described above were going on for millions of years before *man* entered the scene; they will probably continue for eons after we depart. We have a problem—or opportunity—only because people began to help plants grow during the past 7,000 years, a mere speck on the calendar of time.

Plants growing naturally do quite well without our help. They grow in their own comfortable places. Growing slowly, they gather plenty of minerals for balanced nutrition. For all we know, they may utilize thirty or forty active minerals to form perfect cells and tissues. Wild creatures eating this food are well nourished, and they are not prone to heart disease, colds, or cavities in their teeth.

However, people living in towns and cities cannot survive on wild foods or even on the yields of primitive agriculture. We have an artificial society, grossly different from that of primitive people. In this situation we must accept the fact that our food crops and animals will be grown with artificial methods, under rather unnatural conditions.

In gardening it is unnatural to use raw excavation soil for a vegetable-growing area, or to have the garden in the shade of a house or even in the fierce reflected heat of a basement wall. In nature self-respecting vegetable plants will not grow in such places. They leave them for a few centuries of growing cheat grass, cockleburs, and dandelions before trying to grow there.

As a new gardener you would do well to be aware of your temporary assignment as a supervisor of nature, as an assistant in making plants grow. One of the qualifications is to think like a plant. Be leisurely and face the realities of life. If the soil is not ready for carrots and cucumbers this year, dig it up, add organic matter, and get it in shape for next year.

Utilizing the little bit you know about how plants grow, you may follow these general guides:

1. Grow your garden in a sunny place so that the leaves can use ample sunlight in making sugar for your crops.

2. Provide organic matter and mineral-rich fertilizer to assure that the vegetable plants—even when growing

rapidly—may have complete assortments of minerals for normal growth, supplying full-flavored, nutritious foods for you.

3. See that your garden gets enough water to carry minerals and nitrogen upward to sites of growth in the plants, and to replace the water lost by transpiration through the leaves.

By learning how plants grow in nature, you can become a successful gardener and food producer, even under the unnatural conditions of present-day America, where three-fourths of all the people live in cities.

FERTILIZER

4.

HOW TO MAKE AND USE GOOD FERTILIZERS

Getting good garden fertilizers is a problem. When you go to most garden stores seeking such a product, the clerk will usually offer a standard 6–10–4 or 5–10–5 grade. Such a fertilizer is neither very poor nor very good. While providing some cheap, quick-acting nutrients, it is unbalanced, lacking sufficient potassium, magnesium, earth minerals, and organic matter. It will fertilize your garden plants in a mediocre fashion, but it will not fulfill your dreams of beautiful vegetables or help you to be a good ecological gardener.

In this chapter we will describe practical fertilizers for use in growing safe, nutritious vegetables and fruits. Knowing the problems in getting good organic materials and garden fertilizers, we will describe many kinds of materials acceptable for home garden use. Then we will build recipes for ecological fertilizers that anyone can make in any community of the United States. Among these various materials and "do-it-yourself" fertilizers we are sure you can find ones to serve your own desires and needs.

WHAT IS AN ECOLOGICAL FERTILIZER?

An ecological fertilizer is a food for plants that imitates nature in its main features and action.

Completeness and Balance

It contains a full assortment of plant nutrients and minerals, not just the old N(nitrogen), P(phosphorus), and

K(potassium) but supplies of the earth minerals that are so essential in growing normal, pest-resistant plants. The various nutrients are provided in good balance; there's not too much nitrogen in relation to other foods.

Gentle, Sustained Feeding

An ecological fertilizer imitates nature by decomposing slowly in the soil, providing gentle, sustained nutrition for your garden plants. Typically, it will offer some immediate foods, plus increments during the coming two or three months.

Harnessing Soil Bacteria

A good garden soil has an abundant population of useful soil bacteria and a number of helpful earthworms. An ecological fertilizer encourages their presence and vigorous activity by providing organic matter and trace minerals needed by these small organisms.

A fertile soil is a live soil. Its bacteria will gather large amounts of plant foods from the air and soil, giving these extra nutrients to your plants and making you a successful gardener.

This also enables you to grow foods without using so much *purchased* energy. It saves petroleum and earns you the badge of "good citizen."

Growing Pest-Resistant Varieties

An ecological fertilizer fosters pest-resistant plants by providing complete nutrition and vigorous growth. The healthy cells are not easily invaded by insects and disease pathogens. As a result, a minimum of poisonous pesticides is needed in your garden, and your neighborhood will be in no danger of contamination.

Recycling Organic Wastes

An ecological fertilizer is one that contains some organic waste materials, thereby being a better fertilizer and also a waste recycler. It helps to solve our country's massive organic waste problems.

WHAT ABOUT STRICTLY ORGANIC FERTILIZERS?

Strictly organic fertilizers and gardening are ideal, if you can get the necessary materials at reasonable cost, and if you live where you can make compost or obtain good country manure. When America was spread out in farms and rural communities, these conditions were attainable by most people. They could afford to be organic gardeners.

Today it is often quite difficult to follow the strict organic guides. Good organic materials have disappeared from most neighborhoods, and are expensive when available. Bone meal is $40 per hundredweight, five times the price ten years ago. Livestock and poultry are no longer nearby, so getting low-cost manure is not possible for most people. Composting is hard to do unless you have access to waste materials and a place for the compost pile.

Under these conditions we encourage *ecological* gardening as a practical alternative for millions of gardeners, especially those living in cities. It is a way to produce nutritious vegetables with materials available in any neighborhood.

ACCEPTABLE FERTILIZER MATERIALS

In ecological gardening we do not reject any staple fertilizer material because it is chemical or commercial; everything is chemical, even goat's milk, compost, and our own bodies; and nearly everything is commercial. We are interested, rather, in the end results of using various kinds of fertilizers and in the combinations of materials useful to produce safe, fully nutritious garden crops.

Organic Materials

To get good results, a use of some organic materials is essential—the more the better. Dehydrated compost, manure, mulch, seed meals, and bone meal may be purchased in bulk from many garden markets and conveniently used as needed from storage bags. Guides are provided as to how much to use. Some ingenious gardeners obtain good results and save money by using "wet" organic fertilizers. We know of one such gardener who runs vegetable peelings, table scraps, coffee grounds, apple cores, and many other kinds of organic wastes through a kitchen blender, making a glorious mash for use as a garden fertilizer. He applies it to the soil any time of year, according to his own system.

Some gardeners have access to "wet" manures, cannery wastes, compost, and old straw. As general guides for using these materials, let us remember that good farmers use from 12 to 15 tons of dairy manure per acre of land (60 to 75 pounds per 100 square feet of soil). If your garden is about 400 square feet in size, you might sensibly apply from 250 to 300 pounds of "wet" manure. With poultry manure, which is more concentrated, use only half as much, unless the mass is at least 50 percent litter (peat moss, straw, or shavings); then use the same quantity as dairy manure.

In the case of nonanimal wastes, such as "wet" leaves, grass cuttings, cannery wastes, sawdust, peat, or garbage, it is a "weaker" fertilizer and there is little danger that you might burn the plants. However, two other factors should be kept in mind:

- The mass of vegetable matter will attract and "tie up" plant foods from your soil during the time of rotting—to feed the bacteria that cause decomposition—and this process may reduce fertility for a time. Therefore, apply these bulk materials well ahead of spring season, and add a bit of fertilizer to them to hasten rotting. They will repay your efforts in the garden growing season.
- If you apply a large mass of fluffy vegetable materials, you may make the soil so fluffy it will dry out. Avoid this by generous irrigation or by applying such materials well ahead of the gardening season, so they can settle down.

When using organic materials, do not hesitate to mix them. If you have a bit of manure, compost, sawdust, and garbage wastes, go ahead and use them together. The animal kinds help in efficient use of the vegetable kinds. Use common sense in total amount of all kinds you apply.

Dozens of kinds of dried organic fertilizers exist. Most are unavailable to home gardeners; some are badly polluted, therefore unacceptable; some are too expensive. We suggest that you choose, if possible, from the following kinds:

Compost. This is decomposed (rotted) vegetable and animal wastes. It may include leaves, stalks, garbage, manure, sludge, fish, and other organic debris. You can make compost yourself or buy it at the garden store. It costs from $3 to $5 per 50-pound bag. It may be used at 10 pounds per 100 square feet of garden area or in any other sensible amount. It also may be used as an ingredient in your own fertilizer recipe.

Dry Animal Manure. This is usually dehydrated cow manure or feedlot manure from beef-fattening places. It costs from $2 to $3 per 50-pound bag. It may be used at 5 to 10 pounds per 100 square feet of garden area and as an ingredient in your own fertilizer recipe.

Seaweed Meal. Dried and ground ocean seaweed provides foods about equal to dried manure, plus a remarkable assortment of minerals. The use of seaweed helps plants to resist damage by insects and plant diseases (see Chapter 5). Seaweed meal costs about forty cents a pound in 5- or 10-pound bags. It may be used at about 1 pound per 100 square feet of garden area and in your fertilizer recipes.

Sewage Sludge. This is sterilized and activated municipal sewage sludge. It offers 5 to 7 percent nitrogen and varying amounts of other foods. It is sold under various trade names, such as Milogranite and Kaporganic, at $6 to $8 per 50 pounds. It may be used at 2 or 3 pounds per 100 square feet of garden area and in a mixed fertilizer.

Garden Mulch. This is usually made of composted bark, rotted sawdust, forest waste, or other bulk organic materials. It is useful mainly to add organic matter in a garden program and to mulch various kinds of plants. The cost is from $1.50 to $3 per 50 pounds. Use 5 to 10 pounds per 100 square feet of garden area.

Peat Moss. There are two kinds of peat moss in garden markets—Canadian sphagnum moss and Michigan or local peat. The Canadian material is brown and fluffy; the Michigan type is usually black and more compact. Cost varies widely. These materials are to be used sensibly as aids in lightening

MINERAL ELEMENTS IN DRIED SEAWEED*

ELEMENT	PERCENTAGE	ELEMENT	PERCENTAGE
Aluminum	.193000	Osmium	trace
Antimony	.000142	Palladium	"
Barium	.001276	Platinum	"
Beryllium	trace	Phosphorus	.211000
Bismuth	"	Lead	.000014
Bromin	"	Potassium	1.280000
Cadmium	"	Radium	trace
Calcium	1.904000	Rhodium	"
Cerium	trace	Rubidium	.000005
Boron	.019400	Selenium	.000043
Caesium	trace	Silicon	.164200
Chromium	"	Strontium	.074876
Copper	.000635	Sulfur	1.564200
Chlorine	3.680000	Tellurium	trace
Fluorine	.032650	Thallium	.000293
Gallium	trace	Thorium	trace
Germanium	.000005	Titanium	.000012
Iodine	.062400	Tin	.000006
Indium	trace	Tungsten	.000033
Irridium	"	Uranium	.000004
Lantanum	.000019	Vanadium	.000531
Magnesium	.213000	Zinc	.003516
Manganese	.128500	Zirconium	trace
Mercury	.000190	Iron	.089560
Molybdenum	.001952	Silver	.000004
Nickel	.003500	Sodium	4.180000
Cobalt	.001227	Niobium	trace
Lithium	.000007	Gold	.000006

*Data from Norwegian Seaweed Institute, as reported in *Review Of Seaweed Research*. Research Series No. 76, Clemson University, 1966.

and improving sandy and hard soils. They contain little or no actual plant foods.

Ureaform. This is a synthetic organic fertilizer containing about 38 percent nitrogen in a "slow-release" form. It is sold under trade names of Nitroform and Uramite for from fifty to sixty cents a pound. It may be used as an ingredient in homemade fertilizers or for special gardening purposes when nitrogen is needed.

Organiform. This is an up-to-date organic fertilizer made from sewage and other organic wastes reacted with ureaform to produce a clean granular product. It offers from 16 to 24 percent nitrogen in "slow-release" form, plus the plant foods in the waste material. While unavailable at present in most garden-supply markets, it will soon be appearing, since production is expanding. This is a potentially useful ecological fertilizer and material for homemade mixtures.

Bone Meal. Made from dried sterilized animal bones, this is a wonderfully useful gardening material. It offers about 20 percent phosphate and 30 percent lime (calcium oxide). However, it has become scarce and quite expensive at about $10 to $12 per 25 pounds. It may be used at 1 to 2 pounds per 100 square feet of garden area, as a good fertilizer when planting seeds (or transplanting young plants), or as a deluxe ingredient in homemade fertilizers.

Cottonseed Meal. This is made by grinding cottonseed. It contains 5 to 7 percent nitrogen and minor amounts of other foods. The cost is about ten cents per pound. May be used at 1 to 2 pounds per 100 square feet of area and as an ingredient in homemade fertilizers.

Spent Mushroom Compost. Wherever available, the organic material used in mushroom growing is a superior garden fertilizer, fully equal to a good compost or dehydrated manure. It costs about $5 per 50 pounds. Use in the same fashion as compost.

Major Earth Minerals

The major earth minerals include phosphorus (phosphate), potassium (potash), calcium (lime), and magnesium (magnesia). Sulfur might be included, since plants use lots of it, but urban air and soils usually contain plenty of sulfur. These materials may be found in many garden supply markets.

Rock Phosphate. This is pulverized phosphate rocks, right out of the mountains. It contains insoluble phosphate, available to some plants and unavailable to others. Using it with manure, or putting it in the compost pile, improves its availability to plants. Rock phosphate costs from $2.50 to $4 per 50 pounds. It may be used at from 3 to 5 pounds per 100 square feet of garden area, along with compost or manure, or as an ingredient in homemade organic fertilizers.

Superphosphate. This is pulverized phosphate rock reacted with sulfuric acid, or phosphoric acid, to increase the availability of the phosphate. It contains 20 percent phosphate when made with sulfuric acid, and 45 percent when made with phosphoric acid. As a concentrated fertilizer, superphosphate *might* be misused. It costs from $4 to $6 per 25-pound bag. Single superphosphate (20 percent grade) may be used at 1 to 2 pounds per 100 square feet of garden area or as an ingredient in homemade fertilizers.

Sulfate of Potash. This is a fertilizer salt, containing about 50 percent potash and 25 percent sulfur. As a concentrated material it *might* be misused. Usually available in 5- or 10-pound packages, it may be used by experienced gardeners at ½ to 1 pound per 100 square feet of area or as an ingredient in fertilizers. A companion material, muriate of potash (60 percent potash), is frequently available, though it is less desirable for home gardening, being "hotter" and saltier.

Greensand. Made of a special ocean marl, this is the organic gardener's favored form of potassium. Gentle and nontoxic, it costs from $2.50 to $4 per 50 pounds, when available. It may be used at 2 to 3 pounds per 100 square feet and in homemade organic fertilizers.

Pulverized Granite. This is ground rock, favored by organic farmers and gardeners as a material to replenish soil minerals. It contains 3 to 5 percent potash. It may be used at 5 to 10 pounds per 100 square feet of soil area and in organic fertilizer mixtures.

Sulphate of Potash—Magnesia. Providing both potassium and magnesium, along with sulfur, this mineral offers 22 percent potash and 18 percent magnesium (magnesia). When available, it costs $2 to $3 per 10 pounds. It may be used at ½ to 1 pound per 100 square feet or as a useful ingredient in homemade fertilizers. Its trade name is Sul-po-mag.

Epsom Salts. This is magnesium sulfate, offering about 18 percent magnesium (magnesia). All plants need this food, which may be used to supplement fertilizers. It's also useful as an ingredient in homemade fertilizers.

Lime. This is pulverized limestone rock (calcium carbonate), although hydrated lime may also be used. It corrects acidity of soils and also provides calcium as a plant food. Costs from $1.50 to $2.50 per 50 pounds. It may be used at from 5 to 10 pounds per 100 square feet, depending on the acidity and texture of your soil.

Dolomite. This is a special form of limestone containing magnesium (calcium-magnesium carbonate). It is a superior liming material, since it contains this extra food. It costs about two dollars per 50 pounds. Use at the same rate as regular lime.

Wood Ashes. This is a useful fertilizer material which may be obtained from a fireplace, bonfire, furnace, or other common sources. The ashes contain 3 to 7 percent potash as well as other minerals in minor amounts. Use 1 to 3 pounds per 100 square feet. It may be added to the compost pile in conservative amounts.

Minor Earth Minerals

At least six trace minerals are useful in growing nutritious, full-flavored garden crops—iron, zinc, manganese, copper, boron, and molybdenum. Other minerals may have roles in helping plants to resist attacks by insects and diseases. In good gardening it cannot be assumed that the soil will automatically provide all the minerals that are so useful in growing nutritious, pest-free vegetables. Organic fertilizers help because they contain minerals gathered by plants in forming their tissues during previous growth. In addition, home gardeners may use the following trace mineral fertilizers.

GUIDE FOR LIMING YOUR SOIL

Liming the soil has two main purposes—to correct excess acidity and to provide calcium as a mineral food.

Acidity is measured in pH units, which indicate the amount of hydrogen ions that are "loose" in the soil. A pH of 7.0 is neutral, being the acid-alkaline balance in pure rainwater or distilled water. Smaller numbers, such as pH 6.0, 5.0, and 4.0, indicate increasing acidity. Higher numbers, such as pH 8.0, 9.0, and 10.0, indicate increasing alkalinity.

For comparison, diluted hydrochloric acid or vinegar (acetic acid) might have a pH of 4.0, while a laundry bleach could have an alkalinity of pH 10.0. A good garden soil would have a reading of pH 6.4 to 6.8—quite near to neutral.

Centuries of high rainfall removes lime and other minerals from the soil, so it is not surprising that rainy areas usually have rather acidic soils, from pH 4.5 to 6.5. On the other hand, dry regions with annual rainfall less than 20 inches develop alkaline soils with pH ranges from 7.0 to 8.5 or more.

A soil analysis is desirable for gardening; it will show the pH and lime requirement. However, in the absence of a soil analysis you may use the following guide:

	APPLICATION OF LIME OR DOLOMITE PER 100 SQUARE FEET OF GARDEN AREA	
ANNUAL RAINFALL OF YOUR AREA	SANDY SOIL	AVERAGE LOAMY SOIL
Below 20 inches	None	None
20 to 40 inches	3 lbs.	5 lbs.
Over 40 inches	5 lbs.	10 lbs.

Apply the lime or dolomite before plowing, spading, or tilling the garden area, so that it will be mixed with the top 6 inches of soil.

Fritted Trace Elements (FTE). This material provides iron, zinc, manganese, copper, boron, and molybdenum in a gentle, slow-release form. FTE is not widely offered in garden stores, but if available it is a superior material for use in growing high-quality mineralized foods. It may be used at about 4 ounces per 100 square feet of garden area, on the compost pile, or as an ingredient in homemade fertilizers.

Chelated Trace Elements. This is a product in which the particular mineral—iron, for example—is provided in an organic molecule. Many garden stores offer chelated minerals for sale. They should be used according to directions.

Seaweed (Again). Seaweed meal and liquefied seaweed are useful materials to provide whole assortments of minor minerals for gardening use. These products are widely available, although telephoning may be necessary to locate supplies in your community. The seaweed meal may be used as an ingredient in homemade fertilizers.

Pulverized Rock (Again). Granite dust and other kinds of pulverized rock are favored by organic gardeners as materials to restore mineral resources in soils. Compost and manure help to make the minerals in pulverized rock available to plants.

Liquid Fertilizers

Liquid fertilizers are miracle-makers in home gardening. Easy to use, they nourish plants through leaves and stems as well as roots. The foods in the liquid fertilizers are efficient and immediately available to the plants. The following kinds may be found in garden stores:

Liquid Fish Fertilizer. The kinds most widely available are Atlas and Alaska, both being by-products from the fish meal industry. Other kinds are to be found in local markets. Use as directed for supplemental feeding of garden plants.

Liquefied Seaweed. This is seaweed tissue in liquid form, to be diluted with water for spray application to plants or for watering root areas. It has a dual purpose: To feed and mineralize vegetable and fruit crops, and to increase their resistance to insects and plant diseases. This product is widely available and will soon be offered in most garden stores. Use as directed.

Regular Liquid Fertilizers. These products, usually made of soluble fertilizer salts, are available in liquid concentrate form. Mainly, they offer nitrogen, phosphorus, and potash. Use them as directed.

Soluble Fertilizers. These materials are usually made of soluble fertilizer salts, ready to be dissolved in water for liquid application to plants. They are useful in the hands of thoughtful and experienced gardeners.

RECIPES FOR GOOD ECOLOGICAL FERTILIZERS

As a practical matter, we will base several of the following recipes on garden fertilizers already available in your area. This will save hunting for staple ingredients, while still assuring good quality and balance in your home-mixed product.

RECIPE No. 1
ORGANIC-BASED ECOLOGICAL FERTILIZER
Nickname: Eco-Grow

20 lbs. Any good organic-based garden fertilizer available in your area. Rose or tomato fertilizers are acceptable. May be 5–10–5, 5–8–5, 5–10–10, or similar grades.
Estimated cost: $2.50

10 lbs. Compost or garden mulch

10 lbs. Dry manure or mushroom manure

10 lbs. Seaweed meal

50 lbs. Total

Pour these ingredients in a pile on a driveway, street, patio, or basement floor. Mix with shovel and rebag for garden use. Use at rate of 5 pounds per 100 square feet of garden area, tilled into the soil. Also use under rows and hills when planting seeds or transplants, and for side-dressing.

Cost for 50 lbs.: About $8.00

RECIPE No. 2
DELUXE ECOLOGICAL FERTILIZER
Nickname: Deluxe Eco-Grow

15 lbs.	Any good organic-based garden fertilizer as described in Recipe No. 1.
5 lbs.	Ureaform or organiform
5 lbs.	Bone meal (or superphosphate, if bone meal is unavailable)
4 lbs.	Sulfate of potash
1 lb.	Epsom salts (or Sul-po-mag)
10 lbs.	Seaweed meal
10 lbs.	Compost or dry manure
50 lbs.	Total

Pour the ingredients in a pile, as above, and mix thoroughly. If needed to increase bulk or reduce dust, add 5 pounds of slightly damp peat moss or garden mulch. Use at rate of 5 pounds per 100 square feet, tilled into soil. For planting and side-dressing, see description under Recipe No. 1.

Cost for 50 lbs.: About $15.00

RECIPE No. 3
FULL ORGANIC ECOLOGICAL FERTILIZER
Nickname: Eco-Organic

20 lbs.	Any good 100-percent garden fertilizer available in your area
10 lbs.	Ground rock phosphate
10 lbs.	Compost or dry manure
10 lbs.	Seaweed meal
50 lbs.	Total

Pour ingredients in a pile as above, and mix thoroughly. Use as described under Recipe No. 1.

Cost for 50 lbs.: About $8.00

RECIPE No. 4
GARDEN MINERALIZER

Most garden fertilizers commonly available are unbalanced due to a lack of potash, magnesium, and trace minerals. This mixture may be used to correct these shortcomings.

10 lbs.	Compost or garden mulch
10 lbs.	Seaweed meal
2 lbs.	Fritted trace elements (FTE)
3 lbs.	Epsom salts or Sul-po-mag
25 lbs.	Total

Pour ingredients in a pile and mix, as described under Recipe No. 1. Use at rate of 2 pounds per 100 square feet of garden area, tilled into the soil, and as a side-dressing for plants. Use leftover quantity for shrubs, trees, and flower beds.

Cost for 25 lbs.: About $8.00

Many other good fertilizer combinations can be made with various materials described in this chapter. There are five key ideas to use in making your own ecological fertilizers.

1. Use readily available local materials to save costs and delays in planting your garden.

2. Use at least 50 percent organic matter, including good quality manure and compost, if possible.

3. Balance the fertilizer by adding potash, magnesium, and trace minerals, if possible.

4. Use seaweed, if it is available, to supply potash and a whole assortment of minerals and to reduce insect and disease problems.

5. But . . . if you are short of cash, be a scavenger. Gather whatever wastes and fertilizers you can lay your hands on, spade them into your soil, and plant your garden. Good food is going to be scarce and high-priced for some time to come.

When we were young, the barber in our town was named Ben Reimers. Ben's house and big yard were right in the middle of town across from the barbershop. This was a farming community, where many chickens and turkeys were raised. At turkey-killing time for Thanksgiving markets, Ben Reimers closed the barber shop early and gathered loads of turkey feathers for his big garden. We thought he was crazy, and what a mess! Feathers all over town. But next spring, what a garden!

Thirty years later we learned that feathers are 12 percent nitrogen (dry basis) and a wonderful all-around organic fertilizer, rich in the dozens of minerals required to grow foods.

The lesson is this: gather wastes and recycle them in your garden. Be an active ecologist, helping the wastes to get back into America's food-production systems.

Summarizing the suggestions of this chapter, here is a step-by-step outline of a good program for fertilizing your garden. If you have only a few plants or a minigarden, adjust the suggestions for your special needs. If you do not have time for much soil preparation, use common sense and go ahead with your garden. Next year you'll find the time to improve your soil for next spring.

STEP-BY-STEP OUTLINE OF A GOOD FERTILIZING PROGRAM

Soil Preparation

Apply suggested amounts of compost, leaves, manure, garden mulch, peat moss, mushroom compost, or any other bulk organic material that may be available.

Using the mixed garden fertilizer of your choice, add 3 to 5 pounds per 100 square feet of area, broadcast on top of the soil.

Add lime or dolomite at the suggested rate for your soil and climate (rainfall).

If you are an organic gardener, add rock phosphate and/or granite dust to your garden at 5 to 10 pounds per 100 square feet. (If you do this, you may wish to omit the mixed garden fertilizer.)

Then spade or till all these fertilizers and soil improvers into your soil, mixing them to a depth of 6–8 inches.

Sing, whistle, and talk to your garden while it digests all these useful foods.

At Planting Time

Till and cultivate your garden again to get the soil in shape for planting. Add a bit more fertilizer when doing this, if you wish high yields and/or if more than one month has elapsed since Soil Preparation.

Make furrows for the seeds an inch deeper than usual. Apply your mixed garden fertilizer, sprinkled in the furrows (under the rows of seeds) at the rate of about 1 pound per 30 feet of row. Then cover the fertilizer, restoring the furrows to normal depth, and plant the seeds. The roots of the young plants will find this strategically placed fertilizer and thank you for it with good growth.

Use a similar procedure when planting hills of seeds (such as squash and cucumbers) and transplants (such as tomatoes and peppers). Sprinkle a small amount of fertilizer in this soil, too, under and beside the plants, mixed with the dirt.

Side-Dressing

As your garden grows, you can improve yields and quality by sprinkling small amounts of mixed fertilizer (or the mineralizer) alongside the rows or near the plants. Use about the same amount as at planting time—1 pound per 30 feet of row.

But be sensible. Do this only one or two times for long-growing crops (such as tomatoes, peppers, or corn), and as you find useful in your own gardening program and climate.

Mix this extra fertilizer gently into the soil near the plants, but do not disturb the roots. Then irrigate to activate the fertilizer and help in carrying its foods into the root-feeding areas.

Using Liquid Fertilizers

You may use liquid fertilizers instead of side-dressing, if you like. Fish and seaweed are preferred. Use and dilute them according to directions, sprayed on the foliage and allowed to drip on the soil. Or add them to the water used to irrigate the garden plants.

A trombone or other small sprayer is used for applying

liquid fertilizers. For many kinds of plants you may use a sprinkler can. Just add the liquid concentrate according to directions to the water in the can, and sprinkle on the foliage and soil.

This is also the way to apply liquefied seaweed for pest-control purposes. It is done *ahead of pest damage time* by spraying the seaweed on the foliage of the plants. Two or three applications are suggested, at two-week intervals for many kinds of plants.

You get multiple benefits from use of seaweed and fish fertilizers: (1) a measure of pest control, for certain kinds of insects and plant diseases (see Chapter 6); (2) an increase in yields of the plants and crops; and (3) mineralization of the foods, increasing their nutritional values.

Are these amounts of fertilizer excessive for home gardeners? Technically, they are good amounts—about the same as used by market gardeners for many years. But your pocketbook and personal views are also important. If you cannot afford all these fertilizers, be of good cheer. Friendly birds and rabbits may come along to help the situation, leaving their contributions.

Seaweed
Spray

5.

PROTECTING PLANTS AGAINST BUGS AND DISEASES

The big tomato plant shown in Chapter 1 was never seriously threatened by insects and plant diseases for four major reasons:

1. It was fed with a balanced, mineral-rich fertilizer, including seaweed, so that it had healthy, pest-resistant tissues.

2. It grew fast, quickly outgrowing the stage when flea beetles and other insects could successfully attack it.

3. It is of a virus-resistant strain of tomatoes not easily susceptible to diseases.

4. It is part of a diversified ecological system rather than being one plant among two million others on a big tomato farm, where insects and disease organisms can multiply in overwhelming numbers to attack it.

Home gardeners can utilize these kinds of safeguards for their plants, reducing nearly to zero their needs for toxic pesticides.

INSECT AND DISEASE DAMAGE

In further explanation of the nature of the garden pest problem, let us examine two situations in which insect and disease damage are almost certain to occur.

When Plants Are Poorly Fertilized

Wild plants may not get much food when they grow on poor and rocky soils, for example, but at least they may receive many assorted minerals in fairly well-balanced amounts. Thus, they have a chance to grow normal, pest-resistant cells.

In contrast, plants fertilized with the commonly available 6–10–4, 5–10–5, and other oversimplified garden fertilizers are encouraged to grow soft, imperfect cells that are easily attacked by insects. The water-soluble nitrogen causes such a quick vegetative growth that the plants cannot gather soil minerals fast enough to match this briefly abundant nitrogen supply. Malnutrition occurs, and the plants lose their resistance to insects and diseases.

So once man intervenes with an artificial fertilizer, the evidence shows he had better provide a complete and balanced kind. Otherwise, he may foster a pest problem.

When Plants Are Grown in Monocultures

In the United States monocultures are on the increase. There are big agribusiness fields of a single crop and huge colonies of chickens or cows in one place. For example, 5,000 acres of corn may be planted row by row in a single operation, adjoining another 5,000 acres of corn right down the highway. Or there may be 500 acres of tomatoes planted in a single place. Two million chickens may be growing in one colony, or 100,000 beef cattle in a big feedlot.

In these situations every beetle, grub, germ, virus, fungus, bug, heel fly, botfly, and butterfly has a chance to have a population explosion. With natural defenses removed, toxic chemicals are the only means of pest control.

Luckily, home gardeners have a good chance for success in pest control using ecological methods because garden plots are scattered all over the town and countryside. If potato bugs visit you, they will need a road map; and if they get there, a home garden is so small that you can use the old pioneer's method of control and pick the bugs off into a tin can.

There are, of course, up-to-date ways to cope with garden insects. We will describe them in sufficient detail so you may use some of these methods in your ecological garden.

USING SEAWEED (KELP) FOR PEST CONTROL

Farmers and gardeners have known for many centuries that seaweed is helpful in growing vegetables and other crops. Romans used it to fertilize their crops nineteen hundred years ago. Scottish, English, and Irish people have used seaweed in farming and vegetable growing for at least 500 years.* Portuguese market gardeners on the Massachusetts coast gathered seaweed for growing vegetables a hundred years ago, and the practice still persists in many seacoast neighborhoods.

Along the way, farmers and gardeners using seaweed have noticed that their plants were often exceptionally healthy and free of serious damage by insects. The bugs were not killed by the seaweed used in the field or garden; they simply lost their voracious appetites and their desire to reproduce in massive numbers. Gardens fertilized with seaweed tended to be peaceful places where plants and most kinds of insects accommodated to each other. To be sure, many insects lived there, but without consuming or destroying their native habitat—the garden plants.

Clemson University Seaweed Research

Dr. T. L. Senn, head of the Horticulture Department of Clemson University in South Carolina, began investigating horticultural uses of seaweed in 1959. At that time it was never dreamed that DDT, chlordane, dieldrin, and other toxic pesticides might be banned from use due to pollution of foods and the environment. In earlier work, then, Dr. Senn and his staff studied seaweed as a material for improving the vitality and quality of crops, rather than as an agent for pest control.

Rising public protests against toxic pesticides, along with demands for safe farming and gardening materials, have changed this. In 1973, using a grant of private funds, Clemson University launched the Seaweed Research Project to investigate the use of processed seaweed specifically for pest control. The University of Maryland is participating in this vital research project, under an interuniversity agreement.

*A description of early uses of seaweed is given in *Seaweed in Agriculture and Horticulture* by W. A. Stephenson. Faber & Faber. London. 1968.

The authors are consultants in this seaweed research work, having over twenty years of experience in the uses of seaweed and fish materials in growing pest-free plants and crops. The Seaweed Research Project is currently supported by funds from the Economic Development Administration of the U.S. Department of Commerce and the State of South Carolina.

Glen Graber's Big Vegetable Farm

As the proprietor of Graber Produce Company, Glen Graber cultivates 500 acres of fertile Ohio soil that has been in his family for over sixty years. His father cleared the land, near Akron, in 1911. Graber produces all kinds of vegetables —corn, peas, beans, cucumbers, radishes, cabbage, tomatoes, squash, peppers, celery, lettuce, and a dozen more.

In 1960 Glen Graber revolted, personally, against poisonous pesticides. As he says, "For years, one of my small pleasures was picking a sprig of celery, a lettuce leaf, or some other morsel and savoring its familiar taste as I walked through the fields. Along about 1960 I became afraid to do this because of the poisons I was using to grow the crops. It was a very personal thing. And I said, 'If I am afraid of these vegetables, how can I ask other people to eat them?' "

Hearing of Dr. T. L. Senn's work, Glen Graber visited Clemson and commenced his personal and professional relationship with "T" Senn and his research staff. Graber's big farm soon became an unofficial research laboratory in uses of seaweed for pest control.

By 1966 Glen Graber and Dr. Senn had attained sufficient knowledge of the science and art of seaweed farming so that Graber was able to discontinue the use of other means of pest control. Since then he has grown 400 to 500 acres of pest-free vegetables a year for premium markets, with seaweed as the main crop protectant.

The Principle —Why Does Seaweed Work?

We believe seaweed provides a measure of plant protection because it supplies a full assortment of minerals that are active in the life systems of plants. They, in turn, may use these minerals fully to mobilize their enzymes and hormones in natural defenses against insects and disease organisms and to form strong, healthy tissues that resist damage by insects.

Since a plant can attract bees and other beneficial insects by use of its enzymes and hormones, we believe it may have the power to repel unwanted invaders.

In other words, "modern" farming and gardening, with its chemical means of fertilizing crops, has tended to disarm the plants, removing their natural defense mechanisms. The mineral-rich seaweed helps to restore natural defenses against insects and diseases.

How to Use Seaweed for Pest Control

Official research in seaweed for pest control is just beginning in America—about ten years late due to usual problems with public agencies. However, the following guides may be used, based on twenty years of practical experience by the authors, on the Clemson research, and on Glen Graber's big success model. The seaweed used for farming and gardening—except that gathered fresh at the seashore—is in two forms:

Seaweed Meal. Ocean seaweed is harvested, dried, and ground into a meal. Most of the American supplies come from Scandinavian and other northern countries. The demand is rising, however, and our own seaweed resources will eventually be used in increasing amounts. Seaweed can be grown in bays and estuaries as a cultivated crop.

Liquefied seaweed. This product is made by hydrolyzing and pressure-cooking the seaweed, converting its tissues to a liquefied form. It is available as a liquid concentrate to be diluted with water or as a wettable, "instant" powder to be dissolved with water. In either case the liquefied seaweed is used for spray application to plants or for mixing into irrigation water.

Amounts and Ways to Use Seaweed

For growing garden vegetables, the following guides may be used.

Apply seaweed meal at the rate of 1 pound per 100 square feet of garden area, mixed or tilled into the top 4 to 6 inches of soil. Do this when working up the garden area in springtime. Seaweed meal may be included in a mixed garden fertilizer as described in Chapter 4.

Apply liquefied seaweed to the foliage two or three times during growth of the plants. The diluted seaweed can be sprayed or sprinkled with a watering can to wet the leaves thoroughly. Space the applications ten days or two weeks apart. Start this procedure ahead of the insects, before you expect infestation and damage; this enables the plants to assimilate and use the seaweed substances.

Like any valuable skill, ecological gardening with seaweed has to be learned, and you have to earn success. It took Glen Graber five years to clean up his fields, mineralize the soil, reduce insect populations, and attain full control . . . and he is still learning for particular crops and insects.

This is not a wonder drug or a hypodermic injection. It is part of a whole system of gardening in which you join with nature and supervise some of her processes and miracles.

Costs:

At present prices seaweed meal is about forty cents a pound, and liquefied seaweed is about $1.50 per 8-ounce bottle. A supply of these materials for a garden of 500 square feet will cost about $5.

ORGANIC GARDENING FOR PEST CONTROL

Organic gardening is an alternative way to provide mineral fertilization of plants and crops so that they may resist attacks by insects and diseases. However, the mineral supplies in compost and manure vary in different climates. In a high rainfall area they may not be adequate to afford a full measure of protection.

Plants incorporate various minerals into their tissues, depending on the supplies in soils and fertilizers. A soil percolated for centuries by 40 inches of rainfall will have a low mineral content, and the poverty of the soil will be projected into animals, birds, manure, vegetation, and compost piles. Just being "organic" is then an inadequate remedy for a general low level of mineral resources in an area.

Seaweed or other mineral materials should be included in fertilizing programs of such high rainfall areas, to supplement the good practices of organic gardening. The Rodale organic-gardening group encourages the use of seaweed and has included it among approved organic materials for many years.

FISH FERTILIZERS FOR PEST CONTROL

Indians and other early Americans were quite sophisticated when they fertilized maize with a fish under each plant. The fish was a convenient package of minerals; if from salt water, it surely rivaled seaweed as a complete mineral fertilizer.

In recent years several good fish fertilizers have been offered in garden stores. The authors made Marina, Hi-tide, and Fish n' Six during 1952–1962, using scrap fish and fish wastes as primary ingredients. These superior gardening materials offered a measure of pest control wherever they were used.

There is a need to resume and expand the use of fish wastes and other marine materials in U. S. farming and gardening. These wastes are pollutants of land and water near fish canneries, freezing stations, and fish meal factories. Converted to fertilizers, they may contain 14 percent nitrogen (dry basis), plus all the kinds of minerals in seaweed.

Fish fertilizer is *energy*, more useful in farming and gardening than nitrogen products made from petroleum and natural gas.

We believe fish fertilizers can be made that are as effective as seaweed for pest control. Good combinations of the two are possible, with fish providing high nitrogen content and seaweed yielding potash and trace minerals. This should be a rich ledge for probing by the Environmental Protection Agency (EPA) and the U. S. Department of Agriculture (USDA).

WHAT TO DO IF THE BUGS COME ANYWAY

In the foregoing pages we have outlined basic strategies in ecological gardening to reduce the pest problem to an absolute minimum. But what if certain insects and molds ignore these fine strategies and bite our plants, threatening to destroy them? What if the cutworms chew the young cabbage plants, or if the aphids suddenly get the upper hand on our string beans and peas?

Let us not be like the fine old ecological lady who could think of nothing better than beating her potato bugs to death with a hammer on a block of wood.

Entomologists are making progress with new nontoxic pesticides; they are also resorting to some effective old methods of safe pest control used years ago. One gardener reports that a solution of ground chili peppers and water sprayed on plants helps to repel insects. We suggest you choose from the following list of relatively safe pest control methods and materials, if the pests threaten seriously to damage your garden plants.

To Control Slugs and Snails

Pour stale beer in a shallow pan, such as a pie tin. The beer attracts these invaders as effectively as a poisonous slug bait, and they drown in it. Put the pans at strategic places in your garden.

To Control Cutworms

Cut the bottoms out of paper cups and put them around the stems of cabbage, broccoli, cauliflower, and other plants susceptible to cutworm damage. Or use small bands of metal or plastic to protect the stems at ground level. These measures will serve until the plants get large and tough enough to repel the cutworms.

To Keep Away Aphids and Flea Beetles

Put strips of aluminum foil on the ground underneath such susceptible plants as peas, beans, peppers, tomatoes, and cabbages. Use full width on each side of the rows for this portion of the garden. The reflection of light disturbs these pests so that they may not multiply in dangerous numbers. The foil is not harmful to the soil; it will even give a bit of weed control and conserve moisture.

To Control Corn Ear Worms

Put several drops of mineral oil on the silks of the young ears of corn. Be careful, because too much on the ends of the silks will interfere with pollination.

To Control Flea Beetles on Tomatoes

To keep the beetles away during the critical time when the plants are young, spray with soap and water. Or dust with lime when moisture is on the plants, so it will adhere to the leaves. Put the lime in a porous cloth bag and shake it over the plants.

To Control Caterpillars

Use a new biological pesticide called *bacillus thuringiensis,* which is becoming available in many garden stores. It is a dry powder, easily mixed with water and sprayed on leaves and stems of susceptible plants. It is an antitoxin, safe to use.

To Control Codling Moths and Some Other Insects

Use a botanical material called *ryania,* which is sold in some well-stocked garden-supply places. It is a powder made from roots of a South American plant. Mixed with water and sprayed on the leaves of susceptible trees and garden plants, it causes codling moths, fruit flies, and some other egg-laying kinds of insects to get lazy and quite harmless. Use according to directions.

To Reduce Pest Populations

Apply a dormant oil spray in winter and/or early spring before the leaves come out. Spray bushes, trees, and garden borders. This kills many kinds of hibernating aphids, mites, moths, flies, and scale insects, at the same time destroying their eggs so they cannot hatch. Gardeners who use this practice, along with seaweed and other ecological controls, may have relative freedom from garden pests.

To Reduce Exposure to Diseases and Insects

Use pest-resistant varieties of vegetable seeds and plants. To do otherwise is to miss a vital tool in successful gardening. Plant breeders are adding good selections of resistant varieties each year, and these are clearly shown in seed catalogs and on seed packages. Some years ago the resistant kinds were immune mainly to virus and disease infections; for instance, tomatoes could be resistant to mosaic disease. However, groups of plants are now being developed that resist damage by insects, too; for example, there are kinds of vegetables that resist leaf miners, mites, aphids, and loopers. This is a good trend that may assist ecological gardeners.

To Handle Population Explosions if They Occur

If a serious insect infestation threatens your garden, despite the above gentle controls, use rotenone, pyrethrum, malathion, or a new form of pyrethrum called resmethrin. Rotenone, the old safe gardener's standby, has a very short-term effect. The resmethrin is longer-lasting in its protection of plants and is generally more effective than either rotenone or pyrethrum—if you can find it in garden stores. As a last resort use malathion—carefully. It does not pollute soil, water, or foods if used properly, since it fully decomposes shortly after its time of duty. Malathion is no more dangerous to people and pets than rotenone; however, it is a strong poison that can make you very sick if inhaled or liberally applied to the skin. Be sensible, be careful, and always follow directions.

Successful gardeners should learn to use common sense and their ability to observe the workings of nature. Hating all bugs, bees, worms, flies, and snails is a poor trait that will handicap you as a gardener. Some creatures do little harm, and others are actually helpful. An ecological gardener learns to live with all kinds of plants and creatures while growing a useful crop.

6.

MINIGARDENS

One of the finer scenes in America today is someone spading up a 5-by-8-foot minigarden, planting in a window box, or watering a vegetable garden in pots on the fire escape. What do these people have in common? A dislike for tasteless, rubbery tomatoes at fifty cents a pound and for travel-weary lettuce at forty cents a head—and the enjoyment of creating minigardens.

In a heedless world of pavement and business machines live young plants are a pleasure and a slender promise that life may be better sometime, somewhere, maybe. However, the pleasure dims unless you are able to grow *good* tomatoes and lettuce in your minigarden.

In the present chapter we will provide suggestions for successful growing of vegetables in pots, boxes, and flower beds.

THE OLD GREEN THUMB FRAUD

How many times have you heard someone say, "Oh, Mrs. Jones grows lovely things, but she has a green thumb." The fact is that Mrs. Jones's thumbs are just like yours. Her plants grow because she does a few things right, either accidentally or on purpose.

Plants are quite broad-minded. They grow equally well for anyone who provides them with medium-good soil, air, water, light, and temperature. Having a green thumb is providing these things. Talking to plants is fun, but it cannot replace water or a loamy soil with enough air in it so that plant roots can breathe.

So forget your thumbs and be a sensible provider. The plants will love you even better than they love Mrs. Jones.

WHAT TO GROW IN POTS AND BOXES

Honestly, you can grow nearly anything in pots and boxes that you can grow in a regular garden—even corn and potatoes, if you learn how to do it. Just to expand your mind, remember that the Japanese and Chinese learned centuries ago to grow trees in shallow pots, using bonsai culture. They grow apple, pear, oak, maple, pine, spruce, cherry, orange, and most other kinds of minitrees, sometimes passing a little tree from generation to generation until it becomes over 200 years old.

You want vegetables, not trees, however, and any of the following are suitable in a personal project:

Tomatoes
Bell peppers
Parsley
Garden cress
Thyme
Chives
Turnips
Lettuce
String beans
Radishes
Cucumbers (dwarf kinds)
Zucchini and yellow squash
Spinach
Collards
Carrots
Beets (and beet greens)
Green onions

To be exotic, you may add eggplant, herbs, and strawberries.

If you have a sun porch, patio, or balcony, you can grow all these kinds and more. But if you have only a window area or a fire escape, it will be wiser to start with lettuce, radishes, green onions, and bell peppers.

Pots, Boxes, and Other Containers

To visualize suitable pots and boxes, think of plants in a regular garden. A small variety of tomato plant may grow 3 feet tall and send its roots out 2 feet on each side. Most of the roots are concentrated close to the main stem; the

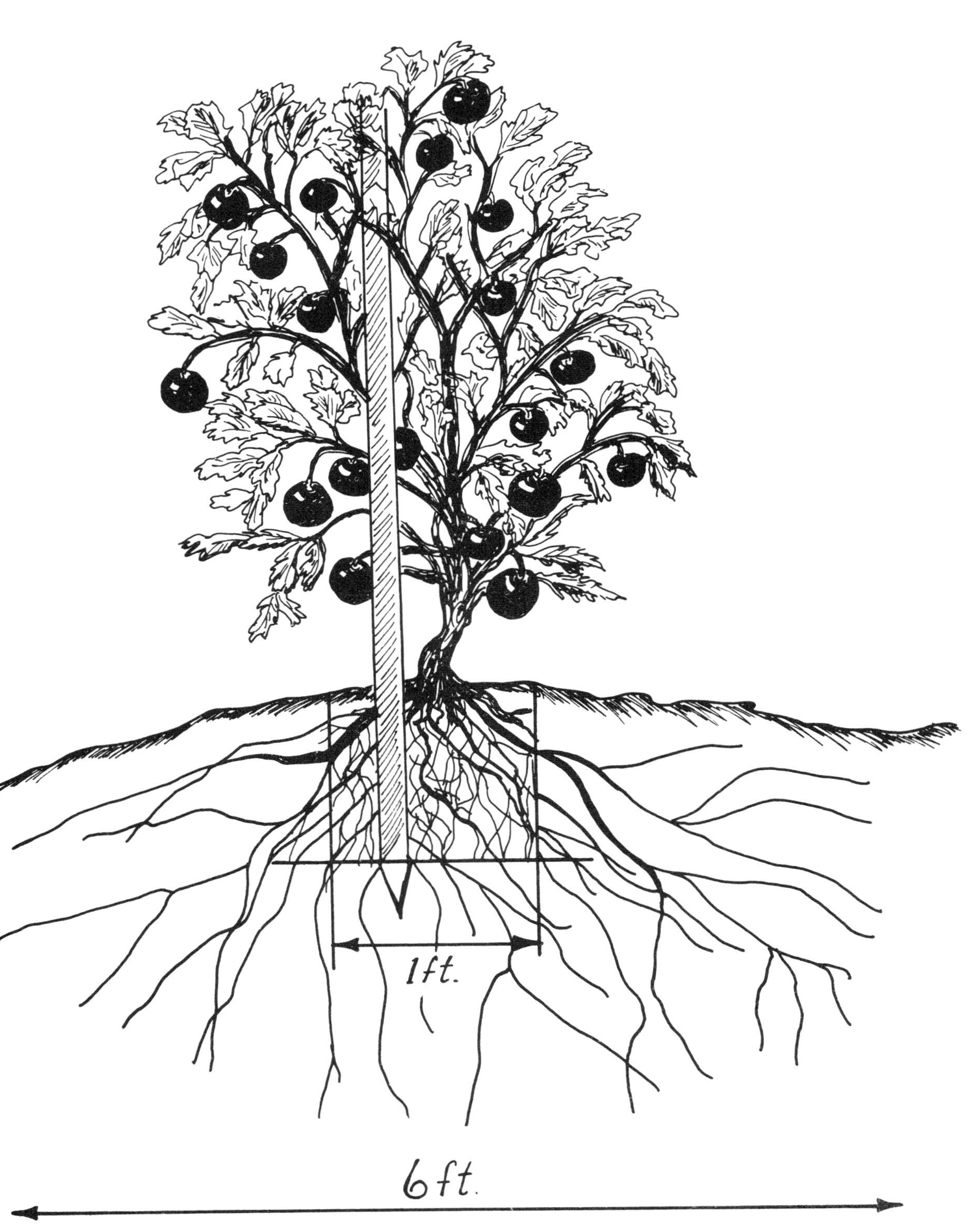
1ft.
6ft.

rest are out there reaching for food and water. If *you* provide some of this food and water, the tomato plant can do quite well in only a small portion of the whole soil area.

The diagram on page 70 illustrates the situation.

Many kinds of containers are available, some of them from previous household use. The accompanying list will assist in your selection of pots and boxes to use in your vegetable-growing projects.

You will learn from experience what sizes of pots and boxes to use. As a starting guide, 10-inch pots, holding a bit over 2 gallons of soil, are suitable for green peppers, tomatoes, cucumbers, and zucchini. Radishes, leaf lettuce, and green onions can be grown in boxes or in most other containers. Carrots need soil at least 8 inches deep to form their long, narrow roots. You can grow several carrot plants in the same container because they do well close together.

If you have room, a raised bed makes a deluxe minigarden. If it is 3 feet wide, 6 or 8 feet long, and 10 inches deep, it can be your main facility for vegetables in rows, supplemented with tomatoes, cucumbers, and squash in individual pots. Window and planter boxes may be used in this way, too.

Remember, soil is heavy. A cubic foot of it in a container may weigh 30 pounds, and the soil in a raised bed or big planter box may weigh 800 pounds or more. Be sure the boxes and containers are strong enough to hold such weights.

Drainage Holes

Clay pots have drainage holes to keep plant roots from getting water-logged. You should punch a few holes in tin cans and other tight containers for the same purpose. Saucers underneath will catch surplus irrigation water; you will soon learn to water the plants just enough so that it seeps through without making a mess. Big wooden boxes and planters have enough extra soil to absorb water, so drainage need not be a problem. Putting gravel in the bottoms of pots and containers is a good practice to facilitate drainage.

SOILS AND FERTILIZERS

Good topsoil is hard to find in cities; also, it is a bit heavy for use "as is" in pots and boxes. Therefore, we sug-

CONTAINERS FOR MINIGARDENS

KIND OF CONTAINER	COMMENTS
Peat pots, made of pressed peat moss, various sizes. Available in garden stores.	Good for starting plants from seed. You transplant the whole pot.
Milk cartons, various sizes.	Useful for medium and small plants.
Used coffee cans, both 1 lb. and 2 lb.	Useful for medium and small plants.
Pressed fiber and pulp pots from nursery and garden stores. Various sizes	Strong and lightweight. Excellent for home vegetable projects.
No. 10 tin cans.	Available from hospitals, restaurants, and nursing homes.
Clay pots, various sizes 2-inch to 14-inch.	Suitable and reliable for many kinds of plants.
Styrofoam pots. Various sizes.	May be useful when you get skilled in using them.
Produce baskets, in bushel and half-bushel sizes.	Good when lined for holding larger plants.
Clean garbage cans, various sizes.	Good for larger plants.
Waste baskets, secondhand or new.	Useful for larger plants.
Plastic bags, various sizes.	May be filled with dirt and used inside outer boxes and pots.
Wooden boxes, tubs, and cut-down barrels.	Make good containers in all sizes needed.

gest that you buy or make the soil for this kind of mini-gardening.

If you have just a few pots and containers, the ready-mixed potting soils are useful. Available at most nurseries and garden stores, they cost about $10 for a bag of 4 cubic feet. This equals about 30 gallons, and will fill a dozen 10-inch containers at a cost of eighty cents each.

A bushel of ready-mixed potting soil costs about $4. It will fill four 10-inch containers at a cost of $1 each.

Mixing Your Own Soils

It is quite a bit cheaper to mix your own soils, and we recommend it for most pot and box gardeners. The following recipe gives the basic directions; adapt it to your own needs and to materials available in your neighborhood.

POTTING SOIL RECIPE

9 cubic feet (⅓ cubic yard), or
Enough for thirty 10-inch pots
Total cost: $14.00 (45¢ per 10-inch pot)

Peat moss, sphagnum type (Canadian) 4 cubic feet (bag or bale)	$7.50
Vermiculite, perlite, or other light aggregate, 4 cu. ft.	5.20
Lime or dolomite, 3 lbs.	.20
Mixed fertilizer, 3 lbs.	.60
Use a good organic and seaweed-based fertilizer as described in Chapter 5, or an organic-based rose or garden fertilizer from a garden store.	
Topsoil, compost or mulch, 1 cu. ft.	.50
Total	$14.00

Pour these materials in a pile on a clean driveway, street, floor, patio, or basement floor; bulk materials first, lime and fertilizer last, on top of the pile. Mix thoroughly with a shovel, dustpan, pie tin, or other handy implement. Sprinkle with water to keep down the dust and to add moisture to the mixture. When mixed, put in moisture-proof bags or boxes until time of use.

This potting soil has enough fertilizer in it to produce good home vegetables. However, for excellent results add a bit of supplemental feeding as the plants grow. Liquid seaweed and fish fertilizers are ideal for this purpose, offering a full assortment of trace minerals to build nutrition and flavor in the vegetable crops. Just add a bit of these liquefied fertilizers to the water used in irrigating the plants. Do this every week or two, depending on the kind of crop.

See Chapter 5 for additional information about fertilizers and supplemental feeding.

SEEDS AND TRANSPLANTS

Planting seeds and watching them grow is half the fun and wonderment of home gardening. If you have children, let them help you. If not, shed off the years and enjoy it yourself.

If there is a good garden store in your area, an experienced clerk can help you to select seeds for minigardening. However, some of the best guides and suggestions are to be found in seed catalogs available from many mailorder companies. Note particularly the dwarf and low-growing varieties the seed companies suggest for use in containers, patios, and other limited spaces.

Sowing the Seeds

Some kinds of seeds can be sown directly into containers—cucumbers, squash, onion sets, lettuce, radish, carrots, beets, and beans. Tomatoes and green peppers are more easily handled by purchasing young plants—if good kinds are available—or growing them for transplanting.

The seed packets give good instructions for planting, telling how deep and when to plant. It is a cardinal rule always to press the soil firmly (but not too hard) on the planted seeds so that they will absorb moisture and not dry out.

Growing Little Transplants

Put sterilized, sandy soil in any convenient container —cut-down milk carton, dish, pot, pan, or shallow box. A fifty-fifty mixture of potting soil and sand is suitable, baked

GOOD VEGETABLE VARIETIES FOR MINIGARDENS

The following suggestions of varieties are based on recommendations from Derek Fell, Director of All-America Selections and National Garden Bureau.

Tomatoes
- Small Fry (All-America winner)
- Tiny Tim
- Hybrid Patio
- Pixie Hybrid

Bell Peppers
- Bell Boy Hybrid (All-America winner)
- Canape
- Italian Sweet

Eggplant
- Morden Midget
- Black Beauty

Lettuce
- Salad Bowl (All-America winner)
- Oak Leaf
- Black Seeded Simpson
- Ruby (All-America winner)
- Buttercrunch (All-America winner)

Spinach
- America (All-America winner)
- Bloomsdale

Radishes
- Cherry Belle (All-America winner)
- Champion (All-America winner)
- Icicle

Onions
- Evergreen White Bunching
- Onion sets from garden store

Carrots
- Little Finger
- Gold Pak (All-America winner)
- Short n' Sweet

Beets
- Ruby Queen (All-America winner)
- Detroit Dark Red
- Golden Beet

Bush Beans
- Tenderpod (All-America winner)
- Romano 14 Italian Bush Bean
- Fordhook 242 Lima Bean (All-America winner)
- Royalty Purple Pod (novelty snap bean)

Cucumbers
- Patio Pik (the only dwarf cucumber recommended for minigardens)

Squash
- Aristocrat—zucchini (All-America winner)
- Chefini—zucchini
- St. Pat—scallop (All-America winner)
- Gold Nugget (All-America winner)
- Summer Crookneck (yellow)
- Table King—bush acorn type (All-America winner)

Strawberries
- Guardian
- Baron Solemacher (alpine strawberry that does not set runners)

These varieties are also suitable for outdoor minigardens in plots and flower beds. We suggest that you use the above list in checking with local advisers and seed stores as to good selections for your particular climate and soil area.

for half an hour in the oven to sterilize it, killing bacteria that might "damp off" or kill tender young plants.

Dampen the soil and sow the seeds in little rows—for larger seeds, four to five per pot, or "hill." Press the soil on the seeds so they will absorb moisture. Cover the container with paper or plastic so that it will retain moisture, and put in a moderately warm place at 60° to 80°. The seeds will sprout in a few days.

After they sprout, put your plants in a place of moderate temperature (50° to 70°) in good light, but not in direct sunshine. When they begin to crowd each other and get "leggy," thin your crop with scissors or by pinching with sharp fingernails. Use common sense to decide how many to thin and how many to leave.

Grow them into sturdy little plants, ready for pots and boxes. Transplant with a kitchen knife or small trowel, dampening the soil in advance so it will adhere to roots. Take a good bit of soil along with the roots to avoid the shock that often accompanies transplanting. If you follow these rules, you'll find that nearly any kind of vegetable plant can be grown from seed and transplanted, often to advantage—even lettuce, carrots, beans, corn, cucumbers, and squash.

LIGHT, HEAT, AND WATER

For most vegetable plants, it is necessary to provide good light and sunshine. Avoid a hot, scorching sun. Imagine you are the plants and you will see how to keep them comfortable. Portable plants may be moved to take advantage of sunshine and to protect them. Stationary plants may be shaded if the sun gets too hot.

Try to provide moderate temperature for most plants, between 50° and 90°. Use sunlight, ventilation, and home heat to regulate temperature. Learn the differences between plants. Lettuce and beets, for example, can thrive in cooler places, while eggplant, tomatoes, and bush beans like warmth.

Water your plants enough to keep them from wilting, but avoid excessive irrigation. Your potting soil has enough peat moss in it to hold its full weight in water. However, the plants remove it constantly, and evaporation also takes a toll. If the moisture content drops below 25 percent of the soil weight, your plants will wilt. Use common sense to avoid this. In warm weather, when plants are growing, they may need water every day. Do not supersaturate them. Enough is enough. Learn the right amount of water through observation and experience.

Minigardeners who wish to explore the realms of artificial heat and light for their plants can obtain information from their County Agricultural Agent. Many cities have Extension Service offices, and hundreds of counties have them in the county seat.

MINIGARDENS IN OUTDOOR PLOTS AND FLOWER BEDS

A good home project is a salad garden. It can be grown in scattered places in flower beds and perennial borders, or against a back fence, or in a plot of spaded soil anywhere.

The National Garden Bureau has provided a model to stimulate thoughts and plans. It is 10 by 15 feet, an area of 150 square feet (see illustration on page 78).

The thirteen vegetables suggested in this model may contribute foods easily worth $100 a season. The seed cost is only $6.30.

The suggestions of Chapter 2, A Home Garden To Save $200, are equally useful here:

- *Sunlight.* Plant your minigarden so it can have sunshine at least half a day.
- *Trees.* Avoid places where tree roots will steal too much moisture and food.
- *Soil Preparation.* Do it well, spading or tilling the soil in advance, if possible, so grass and debris can be decomposed.
- *Fertilizers.* Use organic materials and 4 pounds of a good mixed fertilizer per 100 square feet; use another pound under the rows and plants at planting time.

Gardens in Flower Beds

Some flower beds are gorgeous places for growing vegetables. Fertilized, tilled, and coddled for years, they yearn to grow fine goods.

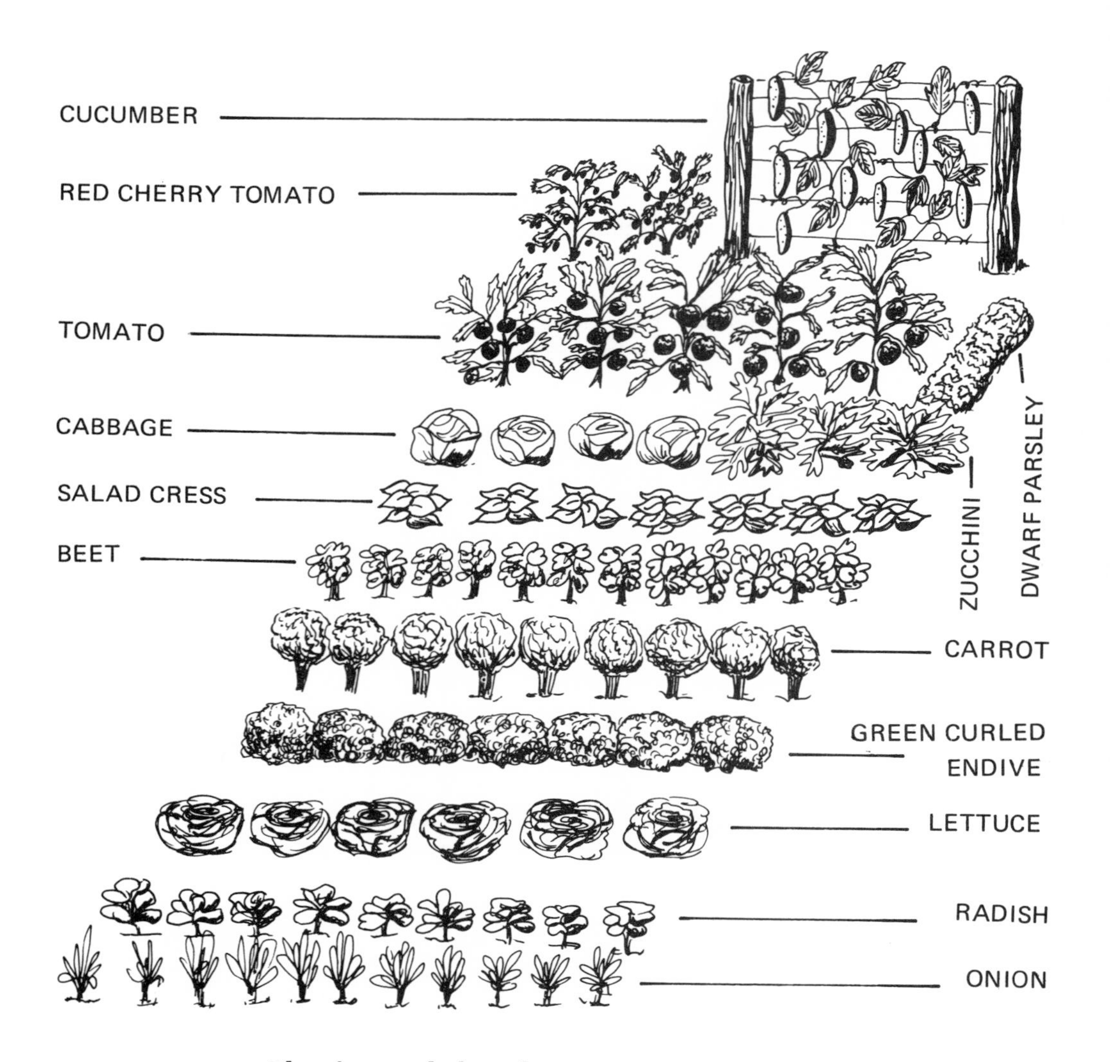

Plan for a salad garden

But other flower beds are horrible. They are made of excavation dirt polluted with mortar, tin cans, and trash and laced with roots of forsythia and other bushes. Such soils can be cleaned up and developed for vegetable growing, but you should not underestimate the task or expect good performance the first year. The treatment has two parts: first, tillage, then adding good, live organic matter. If you add compost, peat moss, mulch, or manure and till it into the soil, a tough flower bed area near the house can often be improved rapidly and made suitable for growing vegetables.

Competition from Shrubs and Bushes

Shrubs and bushes in flower beds affect vegetable plants just as trees affect a larger garden—they tend to steal moisture and food. In some cases the competition makes vegetable growing too difficult. In other areas you can overcome the problem by providing good supplies of fertilizer and water.

The way to find out what will grow successfully in a particular place is to try it. Prepare the soil and plant lettuce and tomatoes, which require only limited root areas. If these vegetables do fairly well, try others. Experience is the best route to success.

Trellises and Doing the Impossible

Several years ago we had a city-bred friend with so little gardening experience that he did not know what he could not do. He bought watermelon seeds and planted them in fertile soil against his house. To his surprise they grew long vines, and having no other place to put them, he trained them up the wall, using string, hooks, and nails. And when the watermelons came, he built a shelf to hold each one.

Which brings us to trellises.

A trellis is a frame to support plants. It can be made by driving two strong stakes into the ground, bracing them, and stringing cords or wires between (see drawing, page 81). Such a trellis can be used to hold cucumbers, butternut squash, or even tomatoes so that their vines occupy minimum garden area.

EDIBLE ORNAMENTALS

People are learning that tomatoes and green peppers look nice in a flower bed. The sharp line between vegetables and ornamentals is fading. And it is interesting to note that a number of standard vegetables of today used to live in flower beds before they were introduced into the kitchen. This is true of eggplants, for example.

As for dual-purpose plants—good to look at and to eat—a new candidate is the rhubarb chard. Its brilliant crimson stalks and the red veins in its reddish green leaves offer color contrasts similar to salvia, only upside down. From time to time the chard can be cropped by the hungry gardener. Remove a few stalks from each plant; more will be ready a week later, and the plants will keep on producing well into late fall.

Ornamental kale—the curly kind—is another plant that is decorative as well as good to eat. Like chard, it can be harvested leaf by leaf and it will produce family greens until December in moderate climates. In the food markets they are worth forty cents a pound.

For its utility and beauty okra should have a place in many gardens. If started indoors, it can be grown in northern zones as well as in the south, but choose a sunny sheltered place. Standing 4 to 6 feet tall, okra bushes display an abundance of red-centered yellow flowers. These give way to edible pods that should be cooked immediately after picking. Fresh, tender okra boiled for 5 to 10 minutes and served with tomato sauce seems a different vegetable from the travel-weary product found in many food markets.

But the grand old sunflower is the ace candidate for new popularity as a dual-purpose plant for food and beauty. Previously grown by farmers to decorate barnyards and provide feed for birds and livestock, the sunflower has a new role due to the unique value of sunflower seeds as human foods. They have an incredible 24 percent protein and 47 percent oil, along with more iron than raisins and more B vitamins than wheat. The unprocessed oil contains vitamin E, lecithin, and all the other hard-to-get oil-soluble nutrients needed by people in a land of over-processed foods.

PLANT TRELLIS

Growing Early Corn to Startle Friends

Corn loves warm weather, and so should not be planted too early in cold soil. However, it can be planted indoors in a big peat pot or cut-down milk carton a whole month ahead of the time for outdoor planting. Kept on a warm porch or near a window, the corn plants will grow 8 or 10 inches high.

They may be transplanted into warm outdoor soil in May, ready to leap ahead making tassels and ripe ears in July. The same procedure can be used in producing string beans and cucumbers. The plants do very well indoors until time for growing in an outdoor minigarden.

HERBS FOR TASTE AND BEAUTY

Many herbs of Old World cultures—Oriental, Indian, Mediterranean, and European—have moved to America. Home gardeners may gain both taste and beauty by including them, and they can be grown in almost any sunny spot.

The varieties can be divided between annuals, to be grown each year from seeds, and perennials or biennials, to be grown in more permanent locations.

The following are annuals; look for their seeds in catalogs and seed racks.

Anise
Sweet basil
Borage
Chervil
Coriander
Dill
Fennel
Marjoram
Parsley
Savory

To grow these herbs, prepare a fertile spot of soil in a sunny place. Add lots of humus and work it in to a depth of 6 inches. Then follow the directions for planting on the seed packages.

The perennial and biennial herbs include the following kinds:

Angelica	Peppermint
Lemon balm	Spearmint
Caraway	Wild marjoram
Catnip	Rosemary
Chives	Sage
Geranium	Tarragon
Lavender	Lovage
	Thyme

Grow these in sunny places in well-fertilized soil. The mint likes quite a bit of moisture, so it may be grown in a rather damp but not too wet place. Some of the perennials and biennials may be purchased from nurseries and garden markets as small plants or rotted cuttings. You may start them from seed and transplant the young plants to your selected places in the garden.

Many herbs* help repel unwanted insects from gardens. The following are some planting suggestions:

Basil	To repel flies and mosquitoes
Borage	To discourage tomato worms
Dill	To trap tomato worms
Mint	To discourage cabbage moths and ants
Rosemary	To deter cabbage moths and certain kinds of beetles
Sage	To repel cabbage moths and carrot flies
Savory	To discourage bean beetles
Thyme	To deter cabbage worms

*Certain flowers also act as insect repellants, among them geraniums, marigolds (nematodes), and nasturtiums (aphids and beetles).

GARDENS
FOR ALL

7.

PLAN FOR A NEIGHBORHOOD GARDEN

The National Garden Survey of 1973 showed that about nineteen million people were interested in neighborhood gardening projects if these would solve their land problems. Even if only half these people might actually take advantage of a community gardening opportunity, it is surely an expression of gardening interest that should not be ignored.

What accounts for this strong tendency among neighbors to plant gardens together? Evidently the central motive, shared by all, is a desire to cope with inflation by saving money in the food budget. Mingled with these practical aims, however, there is surely a rise of old-fashioned community spirit, a desire by people to work with their neighbors on common problems and enjoy a refreshing outdoor activity together. When you garden with your neighbor, you have a sense of town life again, where people call one another by their first names and say "good morning" when meeting on the street. This value in neighborhood life has been missing in many American cities and suburbs.

Neighborhood gardening is expanding in the United States; it is a movement that has had success. In this chapter we will examine existing models and plan a typical neighborhood garden.

HISTORY OF NEIGHBORHOOD AND COMMUNITY GARDENS

Neighborhood gardens are not new; they have a long and successful history in Europe and North America. The early settlers of America often gardened together, and even farmed on mutual holdings. Examples are the Oneida and Amana communities of New York and Iowa and the use of "commons" in New England. A useful variation was the *ejido* property system in Mexico, in which every citizen was assured the right to utilize certain pieces of land for growing subsistence foods.

The Victory Gardens of World War II are another chapter in community gardening in the U.S. The essential land was obtained in public parks, playgrounds, and idle properties, as well as in private holdings loaned for the purpose of growing home food supplies. The Department of Agriculture estimates that about forty million gardeners were growing vegetables in this program in 1944; considering the total U.S. population of that time (about 140 million), this was an impressive participation. Some of the Victory Gardens started then are still in operation today. For example, those of Washington, D.C., are still going strong with about 150 gardeners growing vegetables in plots of about 1,000 square feet each.

Community gardening commenced in Europe in the Middle Ages in part as a measure to alleviate local poverty and hunger. The French evolved three kinds of multiple unit gardens: workers' gardens, near village housing; family gardens, leased from nearby farmers; and industrial gardens, provided by factory owners. Many of them still exist today.

In England land-leasing laws were passed in 1819 to assure that people hard-pressed for food might be able to grow gardens to increase their vegetable supplies. These laws were strengthend by land-allotment regulations under which local authorities could provide land for neighborhood gardens. In more recent years major housing projects in England have included both play areas for children and garden space for families who live in the home units. An example is the big Runcorn housing project near Liverpool, which includes thirty acres of land set aside for gardens.

Derek Fell, Director of the National Garden Bureau, recently toured England and six European countries to study community gardening systems. He reports that the English enterprises in this field are most impressive, with about 57,000 acres of land under cultivation in community gardens, representing a total of 560,000 individual plots. Such gardens are also popular throughout Germany, Holland, and France. An international conference on community gardens is held each year. Food production is featured, but attention is also given to flower beds, garden sculpture, espalier fruit trees, and hedges and esthetic places for relaxing. As Derek Fell says, "Each garden is intended to be a 'little bit of heaven.' "

With this background of success in community and neighborhood gardens during the past 300 years, we may assess United States models that are useful guides for today.

SEATTLE'S P-PATCH

P does not stand for "pea"; it stands for Mr. Picardo of Seattle, who was the previous owner of land used in this community garden program.

As is so often the case, this success model emerged from rather desperate beginnings. It was started in 1971 on a bit of fertile truck-gardening land that was forced out of business by city growth and high taxes. About 150 people made temporary arrangements to use three acres of it, one-third in a communal garden and the rest in individual garden plots.

The project was barely saved in 1972 by an appeal by the gardeners to the Seattle City Council to purchase the land for garden allotments. Instead, the city's Building Department leased the area to assure it would be available for garden use in 1973.

Then, due to its timeliness and popularity, Seattle's P-Patch acquired institutional support that may assure its existence and continued growth. The Park Department plowed and prepared the land with its heavy equipment. The Youth Division of the Human Resources Department acted as public sponsor. And Puget Consumers Cooperative, a nonprofit corporation with over 1,000 family members, stepped in to administer the 1973–1974 program.

Conducted on 300 plots of 400 square feet each, the 1973 P-Patch project was a well-publicized success. It stimulated the City Council to form an Agricultural Action Task Force to propose plans for expansion of the program into a city-wide neighborhood gardening project. By late 1973 over 3,000 gardeners were seeking admission into the P-Patch program.

The USDA Extension Service helped the Task Force locate thirteen sites for the 1974 gardening operations. Ten were city-owned and three were provided by private sources.

The sites included enough land for expansion of Seattle's P-Patch to serve about 1,000 neighborhood gardeners. The overall budget for the year was approximately $25,000.

The gardeners each pay $10 for use of the land. This includes soil preparation, water, and a clean-up tillage at the end of the season. The city is advancing some front-end money, but the sponsors foresee that the P-Patch gardens will attract enough resources to pay their way without tax fund support.

Assistance by municipal agencies during the critical first two years was a major factor in the success of Seattle's neighborhood garden project.

ADOPT-A-LOT IN BALTIMORE

A noteworthy variation in neighborhood gardens is Baltimore's Adopt-A-Lot project. This program emerged in 1973 as a move to help the City's Department of Public Works (DPW) in maintenance of 1,350 vacant lots, many of which had become trash yards, rat jungles, and eyesores.

A local leader named Bonnie J. Buikema proposed that some of these vacant lots should be offered for flower and vegetable gardens. These and other discussions stimulated approval of the Adopt-A-Lot program, in which the city issues a letter of agreement to any qualified "adopter" who will take over the vacant lot, maintain it, and use it.

The approved uses are of many kinds; any constructive proposal may be approved, such as:

- Children's play area
- Neighborhood parking lot
- Flower garden, small park, or rest area
- Neighborhood vegetable garden

Neighborhood gardening on adopted lots began in 1973 with only twelve gardens. These were so successful that a sound expansion occurred in 1974, encouraged by several services provided by city or other agencies.

The Department of Public Works assists by cleaning up the lot before it is transferred to the adopter. If the lot is to be used for gardening, a mixture of leaf mold and sludge is provided. The Bureau of Parks provides wood chips for paths, and it may supply mulching materials and trash containers. The Chesapeake & Potomac phone company has provided garden tools at six locations where gardeners may borrow shovels, rakes, trowels, hoes, hoses, and grass whips to be used in gardening and maintenance work.

Administration of Baltimore's Adopt-A-Lot program is provided by a committee of volunteers. Its role, in part, is one of coordinating various services. For example, it obtains and schedules Agricultural Extension Service visits to gardening groups and acts as a troubleshooter and overseer.

Over sixty lots were adopted in Baltimore in 1974, half of them for neighborhood gardens, and the gardening phase is expanding. The Adopt-A-Lot name and idea seems contagious; it already has been adopted in Syracuse, New York.

Again, this is a nonbudgeted civic program, attracting essential resources and services wherever it can find them. However, the city has incentives to encourage success, since Adopt-A-Lot is a popular way to solve municipal and neighborhood problems. When adequately supported by civic agencies, we predict a wide use and success of the Adopt-A-Lot idea and plan.

GARDENS FOR HANDICAPPED PEOPLE—CINCINNATI

This variation is basic, since every community has its full share of handicapped and elderly people who may derive benefits from gardening.

The Cincinnati project is called Operation Green Thumb. Started in 1970 with only one garden area, it has been sponsored by the Recreation Commission, assisted by the Garden Center of Greater Cincinnati. The 1974 program includes twenty garden areas serving over 500 people, mainly the young. One of the areas caters especially to handicapped children.

Senior citizens are attracted into Green Thumb to assist young people with their experience in gardening. This idea of linking experienced gardeners with young folk is expected to expand.

Cincinnati's garden plots are usually 10 by 10 feet, being smaller than family size since so many are used by children. In addition, each area has a communal garden planted to corn, pumpkins, winter squash, and other large-growing plants.

Using the counsel of the Garden Center, Cincinnati provides a Nature-mobile service as an interesting part of Operation Green Thumb. This educational garden-on-wheels takes easy-to-grow flower and vegetable seeds into inner-city areas to acquaint children with the magic of growing plants.

In this discussion of Cincinnati's gardens for handicapped people we should also remember that blind people may become adept vegetable gardeners as well as students of ornamental horticulture. We remember Walter Thomas, a blind victory gardener near Portland, Oregon, in 1944, who might be weeding and tending his garden at midnight. Since he used other senses than sight, why should he care about the dark or what time it was?

Dr. T. L. Senn of Clemson University reports that blind people often develop outstanding skills in both vegetable gardening and floriculture. To assist in their participation in gardening, he has produced a popular gardening book in Braille. Dr. Senn has also established an attractive blind people's garden on campus property, comprising about two acres. A sign in Braille at the entrance describes this unique resource and tells blind people how to use and enjoy it by feeling and smelling the plants.

Dr. Senn also conducts horticultural projects for young convicts released from penal institutions of South Carolina. By associating with plants and learning skills in gardening and horticulture, they often are able to readjust successfully in society.

It is our belief that gardening and horticulture will have increasing roles serving blind, crippled, retarded, and otherwise handicapped people—as well as millions of low-income families who must have gardens if they are to have adequate diets. Inclusion of these people in neighborhood gardens is one way to open these opportunities, and Cincinnati's gardening projects deserve further study as success models in this field.

GARDENS SPONSORED BY PRIVATE BUSINESS FIRMS—DOW GARDENS

There are several reasons why a bank or business firm might sponsor neighborhood gardens: it promotes good employee relations; it improves the public image; it is good local advertising. Recently, an additional factor has been present, particularly where banks are concerned: gardening helps to sustain the solvency and welfare of the customers, fostering a good business condition in the community.

Several major U.S. companies have sponsored gardens for many years. Dow Chemical Company at Midland, Michigan, is one of these. It started Dow Gardens in 1940, and has sustained them successfully at only minor cost. Users of about 200 plots pay $7 each per year. The company tills the land in spring and stakes out the gardens into 40-by-50-foot areas. Users may request the same plot year after year, to benefit from good fertility and tilth they incorporate into the soil. About 75 percent of the gardens are used by Dow employees and retirees, the rest by other people of the Midland community.

Hercules Corporation at Wilmington, Delaware, is another long-sighted company with an expanding gardening program; however, its projects are mainly for children and young people. A typical Hercules gardening team is ten children and one adult volunteer. The company provides land, spring tillage, fertilizers, tools, plants, and storage facilities. Its Agricultural Division provides several of the adult volunteers.

Rodale Press at Emmaus, Pennsylvania, provides another example of successful company-sponsored gardening —in this case strictly following organic guides. The gardens for employees and other local people are located on the Rodale Experimental Farm, where land is provided for any employee who wishes to use it. These plots are 10 by 30 feet, staked out in a fertile area plowed for all the gardeners in springtime.

FIRST-NAME GARDENS—JACKSONVILLE, ILLINOIS

The term "first-name" was adopted from the idea that gardeners who work together develop a spirit of neighborliness and usually call each other by their first names. It was first used in Council Bluffs, Iowa, where the First National Bank sponsors a community project of about sixteen hundred gardens for its customers and friends.

Max Roegge, President of the First National Bank of Jacksonville, Illinois, visited the garden with other officers of his bank and decided to launch a similar program in his community. Motivated by the bank's desire to encourage solvency in the community and among its customers, the project was established in the spring of 1974.

Jacksonville's "first-name" gardens includes 150 individual plots that are tilled in springtime by the bank and furnished free to customers on a first-come-first-served basis. All of the gardens were taken within six weeks after announcement of the program by the bank. Mayor Milt Hocking of Jacksonville is one of the gardeners.

Where is this phase of community gardening in America going? We believe that we are entering into an era of expansion. The costs to the sponsor are quite low, perhaps $3,000 or $4,000 for the entire year. Where else can a business firm purchase so much good will for this low cost? There is another practical incentive, too, linked to human values. Max Roegge believes the gardens will be especially helpful to young married couples and retired people of the community, enabling them to save $200 to $300 a year per family in food costs. A hundred families doing this add $25,000 of net income in the community, and nothing pleases a banker more than seeing new net income in his market area.

A MODEL NEIGHBORHOOD GARDEN

An assessment of success models in neighborhood gardens shows that the following conditions and factors are desirable, if not essential, to achieve success:

A Strong Local Leader

Nothing quite replaces this. The leadership may be in two people, or even three or four, but it should clearly be present, capable and strong. As a rule, one well-motivated person is the prime factor in starting a successful community gardening project.

Sponsors

Successful gardens usually have one or more strong sponsors. Their roles are to lend civic prestige to the gardening group, to assist with good publicity, to help in gaining access to gardening land, and to help in obtaining low-cost services and resources. In some cases they may provide services and financial aid themselves.

If not present at first, sponsors should soon be acquired. The Seattle P-Patch gardens illustrate this. At the beginning the project was quite frail; one or two adverse incidents might have killed it. Getting assistance and support from the city's Park and Human Resources departments was vitally useful. This helped to win added support from Puget Consumers Cooperative and from the City Council.

Capable local leadership generally attracts good sponsors.

Suitable Land

Access to suitable land is necessary; fortunately, good land is often available even within a city. Mediocre land can be built into fertile gardening soil in two or three years. Experience shows that the gardens should be within walking distance of the gardeners' homes, if possible. Being too far away tends to defeat some of the purposes of the project.

Financing

Actually, this is a secondary problem in a well-conceived project. The gardeners should pay an annual fee of $10 to $20 per plot, depending on incomes in the neighborhood —only $10 or less if incomes are very low. Capable leaders may get services to use in lieu of cash for the more expensive operations. In the examples described you will note that sponsors and public agencies usually plowed or rototilled the land and helped to get it in shape for gardening.

Suitable land can often be obtained free or at very low cost, since it is idle or in public ownership. Parks, arboretums, vacant lots, and public utility properties may often be used. Also, interested private owners may be willing to provide it at low cost. Suitable land can often be obtained free. If leased, the annual cost for fifty gardens should not exceed $50 to $75.

Thefts and Fencing

Theft of crops and damage to the gardens are troublesome problems, especially in city areas where respect for property is at a low ebb. Fencing and guards may be necessary in some places during vegetable harvesting season.

Now let us outline a physical plan for a neighborhood garden, using about one acre of land area. This is 43,560 square feet, an area 208 by 208 feet, about the size of a city block.

Individual gardens of 300 to 400 square feet are good sizes for many persons and families. After allowing for paths and service areas, it is advisable to plan for at least 500 square feet of space per gardener. This means that an acre of land or a big city lot might serve about eighty gardeners, *if* all the land is put into individual gardens.

A good idea if land is available is to block off 20 percent of the area for use in growing vegetables that naturally require a larger growing area—corn, pumpkins, winter squash, melons, and potatoes. If well managed, such larger plantings may serve the entire group of gardeners, or close-knit groups may arrange to use the larger reserved area.

The two alternatives are visualized in the illustrations.

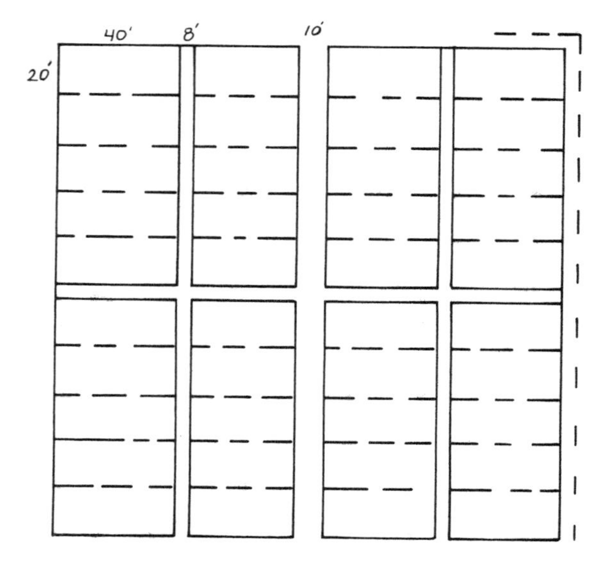

Plan No. 1 All the land is divided into individual plots. In this model a land area about 200 feet square is subdivided into 40 plots 20 by 40 feet in size. This leaves enough space for a center corridor and pathways to reach all the individual gardens, and a margin around the entire gardening area.

Plan For a Neighborhood Garden

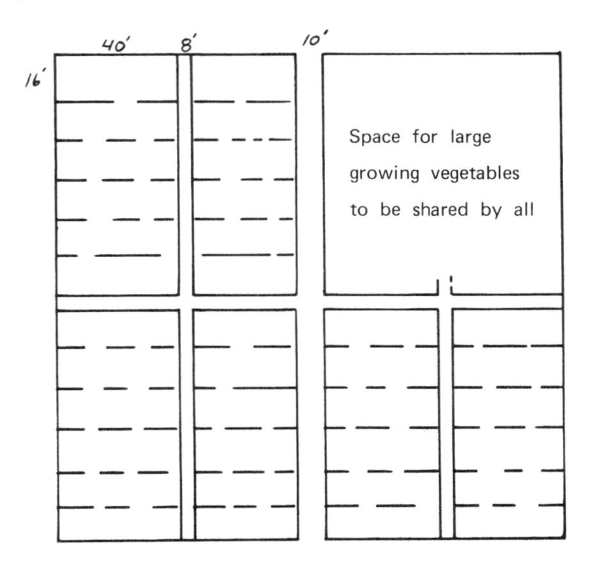

Plan No. 2. Part of the land is reserved for large-growing vegetables. This is a similar land area with 25% reserved for growing such "field" crops as corn, beans, pumpkins, squash, and melons. The individual plots are reduced to 16 by 40 feet in size, since all gardeners may share in the harvest from the "field" crops.

How Big a Neighborhood Will this Garden Serve?

City blocks of usual size often accommodate from twelve to twenty homes for fifty to eighty people. Only some of these residents will wish to participate in a gardening program. Being conservative, we assume that five homes or family groups per city block will elect to take part. Since eighty gardens are available in an acre of land, this means a neighborhood project may serve about sixteen blocks. This is only an estimate for use in preliminary planning; different neighborhoods vary widely in the kinds and numbers of gardens they wish and need.

Neighborhood gardening is on the upswing in America. It is a good and useful trend, probably here to stay. Planners may therefore begin to think in terms of garden needs in building new housing projects, suburban developments, and inner-city urban renewal services. Some new homes can have backyard gardens on their own properties, but apartment and condominium dwellers who have no space will need access to neighborhood gardens within half a mile of their homes.

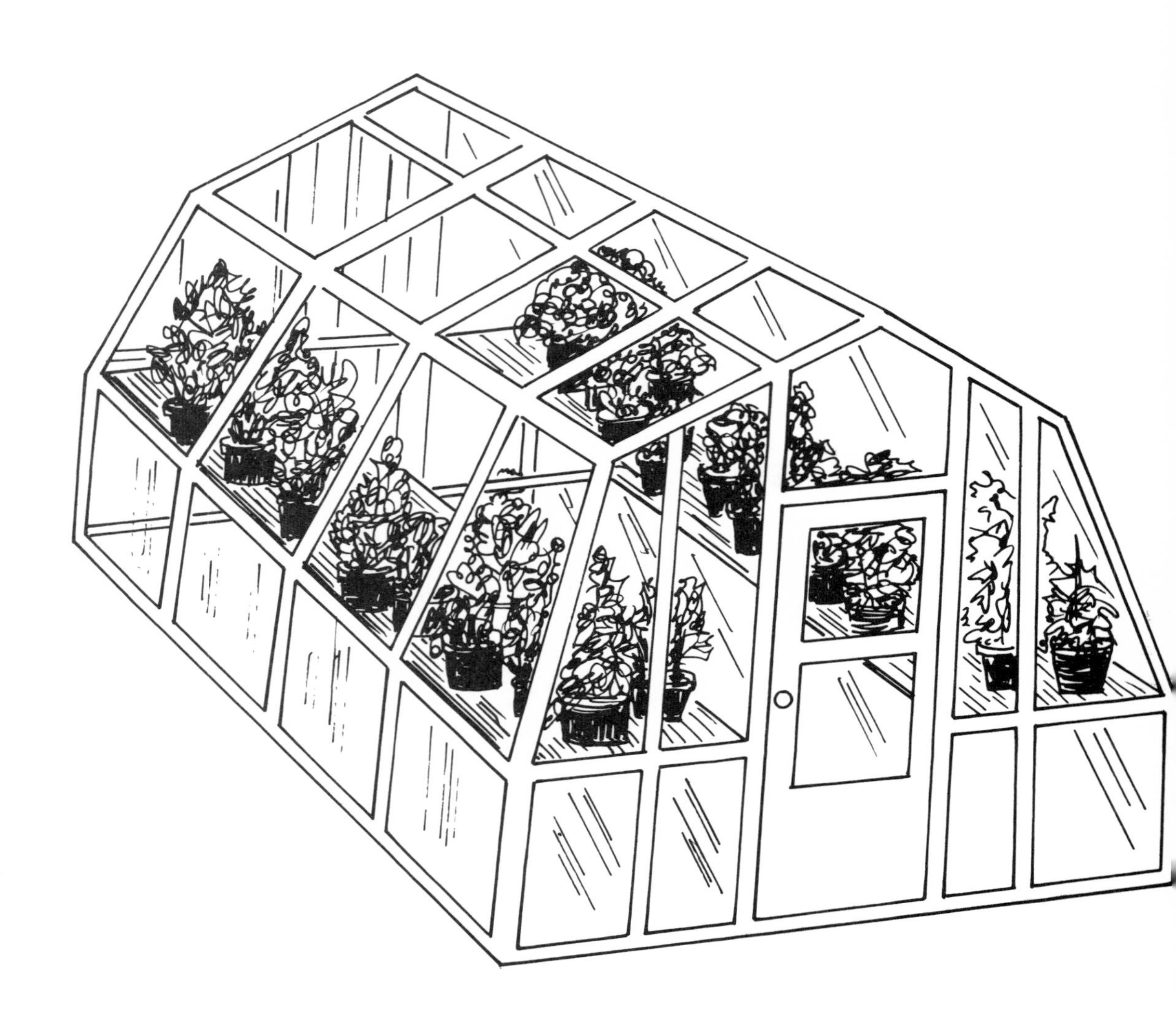

8.

HYDROCULTURE AND HOME GREENHOUSES

What is hydroculture? It is a gardening system that lets you control the temperature, air, nutrition, and water of your plants, producing supercrops. Tomatoes in hydroculture units may yield 300,000 pounds (150 tons) per acre in a single season. This means that a neighborhood unit of 1,000 square feet might produce 6,000 pounds of delicious red tomatoes worth $2,400 at food market prices.

Literally speaking, *hydroculture* means water culture—growing things with water as a main feature of the system. The plants are nourished by means of diluted nutrients in irrigation water. They are grown in soil, but 75 percent or more of the nutrition is provided by foods dribbled to their root areas by perforated hoses. The whole system is enclosed in a low-cost greenhouse.

Hydroponics, on the other hand, is a system in which plants are grown without soil, in a bath of water in which plant foods are dissolved—a dilute nutrient solution circulated by pumps. This system, too, is enclosed in a greenhouse. However, hydroponics is less adaptable for neighborhood use than hydroculture, since it is more costly to install and more difficult to operate successfully.

In the present chapter we will describe a model hydroculture unit suitable for growing vegetables in a local neighborhood. We will also describe small greenhouses adaptable for home production of vegetable crops.

A NEIGHBORHOOD HYDROCULTURE UNIT

This model is suitable for use in any of the following situations:

> By a mutual food production group of about twenty-four families or 100 people, to supply 7,000 to 8,000 pounds of vegetables a year.
>
> As a training and food production facility in a young peoples' gardening program; for example, in the Cleveland or Washington youth garden projects.
>
> As a production unit in a private food production enterprise, growing vegetables for sale.

The costs and annual yields will vary with climate. Boston and Minneapolis are quite different, of course, from Los Angeles or New Orleans. Even the quality of sunlight is different in various places, to say nothing of differences in mean annual temperature. However, this design will serve as an all-around model to illustrate major features of hydroculture gardening in America.

A Neighborhood Group of 100 People

Let us commence at the right place—with a group of people wishing food—and then design a production unit to serve some of their needs.

In this case we will assume a twelve-block area with two families or living groups per block wishing to participate, making twenty-four families. With an average of a bit over four persons per family, this neighborhood group consists of about 100 people of various sizes and ages. If they wish to organize, they may call themselves a mutual food production club or some other suitable name.

Some people like cucumbers, others don't; some eat 2 pounds of fresh vegetables a day if they are available, others nibble 2 ounces. In any event, the families will still buy some of their vegetables at the supermarket, even though they produce home supplies. Taking these factors into account, we will assume that each member of the neighborhood group needs an average of 6 ounces a day of hydroculture vegetables. The neighborhood group of 100 persons therefore needs about 40 pounds of vegetables a day.

Since hydroculture units cannot operate year-round, due to climate and time needed for cleaning and replanting, we will assume that this neighborhood facility will supply vegetables six months a year, or 180 days.

Thus, we should design it to yield about 40 pounds times 180 days, or 7,200 pounds of vegetables to be consumed during the late spring, summer, and fall seasons. In warm and sunny climates the harvest can be extended for another sixty days; however, to be conservative we will base this plan on a six-month calendar period.

Building the Hydroculture Unit

The hydroculture unit on page 102 is 30 by 44 feet—about 1,300 square feet. After providing a work and service area and pathway, the actual vegetable growing portion is 1,100 square feet. The dimensions might be varied; for example, the overall area might be a square 36 by 36 feet, still providing about 1,300 square feet with a 1,100-square-foot growing area. Readers who wish to do so may write to the authors for an engineer's sketch of such a hydroculture facility.

Structure

A low-cost greenhouse structure can be made of wood or metal framing covered with fiberglass. The actual cost will depend on the selection of materials and whether or not full compliance with urban construction codes is necessary. Estimated costs for two different situations are as follows:

Steel Structure (Higher Cost). This steel frame structure, 36 by 36 feet, with a dome-shaped roof 16 feet high, covered with fiberglass skin, includes ventilation and cooling facilities and irrigation equipment. The design will meet construction codes of most cities.

The cost exclusive of labor is about $4.50 per square foot, a total of approximately $5,850.

Erecting this unit will require the labor of three persons for seven days, a total of about 168 person-hours. Preparing the site is a variable item, depending on the land area. We suggest that thirty-two hours be added, making a total of 200 hours (twenty-five person-days) required to build this hydroculture unit.

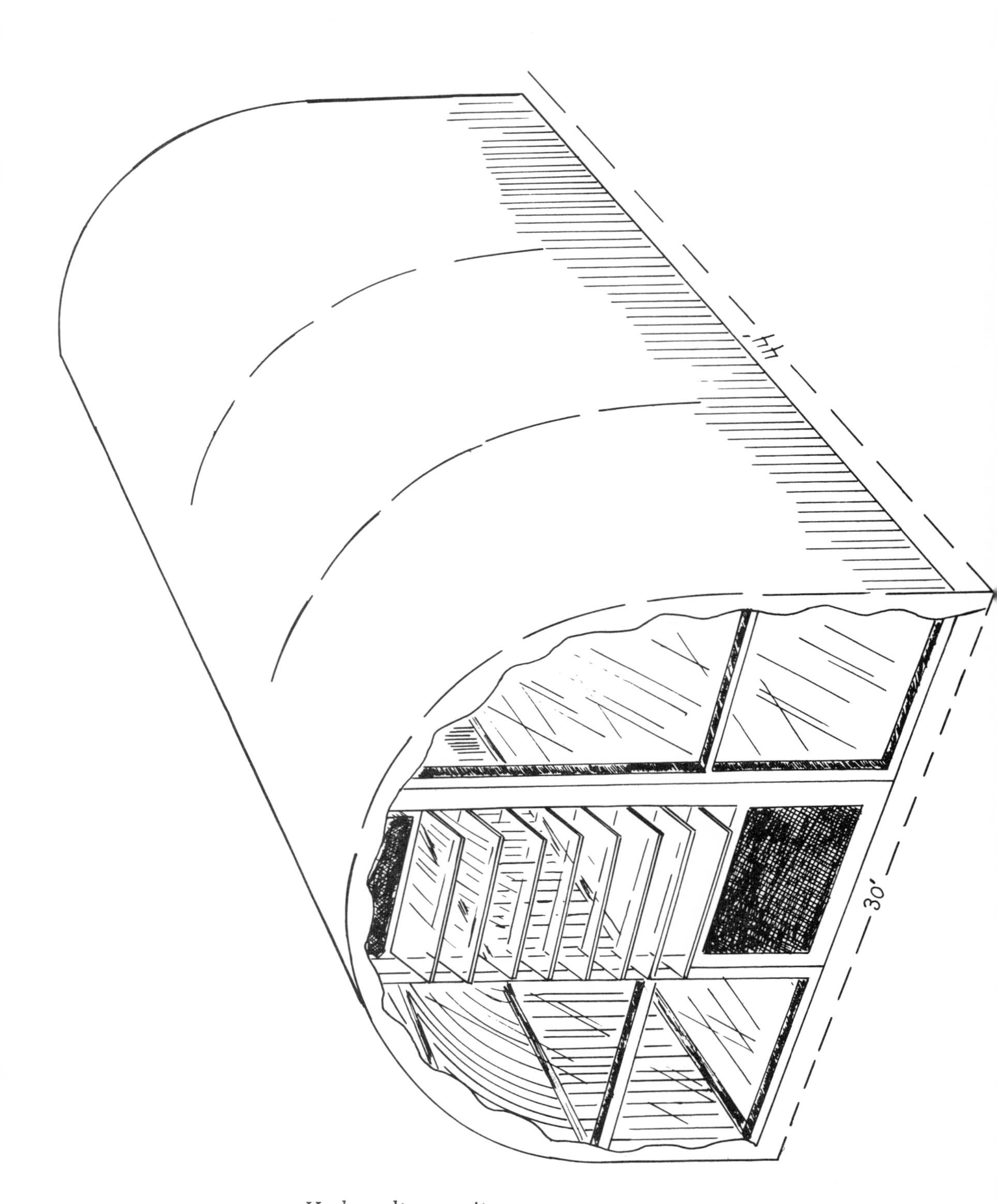

Hydroculture unit

Wood Structure (Lower Cash Cost). This wood frame unit has an A-frame roof covered with fiberglass. Other features, facilities, and equipment are the same as in the higher cost steel frame unit.

The cost exclusive of labor is about $3.50 per square foot, a total of $4,500.

Erecting this unit will require the labor of three persons for nine days, a total of about 216 person-hours. Preparing the site will require the same time as described above, estimated at thirty-two hours, making a total of about 250 hours (thirty-one person-days) to build this lower cost unit.

Land & Soil

The hydroculture unit can be placed on any suitable area, but fertile, well-drained soil is preferable. If good soil is unavailable, however, the unit can be placed on mediocre land, and good top soil and compost hauled in to convert the 1,100 square feet into a fertile soil. In any case the soil is thoroughly sterilized and fertilized before planting (use Chapter 4 as a guide). Good drainage is essential.

The Production System

Such plants as tomatoes and cucumbers are transplanted into rows 3 feet apart, 1.7 feet in the rows, providing 5 to 6 square feet per plant. The production area will then support about 200 to 225 plants. Other kinds of vegetables might be grown with a planting system suited to them.

The crop is grown with a "dribble method" of irrigation and feeding, using special perforated hoses laid alongside the rows. Dilute plant foods are injected into the pipes carrying the irrigation water by means of proportioners that accurately measure the amount to be fed into the stream. The plants are then nourished at the same time they are watered through the hoses.

The plants and crop are supported on trellises. Tomatoes and cucumbers grow 6 to 8 feet high.

Plant Protection Against Insects and Diseases

The ecological principles and pest-control materials described in Chapters 4 and 5 are used. When the plants are well nourished, minimum insect and disease sprays are needed.

Sunlight, Air, and Temperature

These units are heated mainly with sun's heat; therefore, the operations should be adjusted to climate and outdoor temperatures. Fans and good ventilation are used to help in regulating the temperature and quality of air in hydroculture units. Artificial lighting and heat can be provided, if desired, to supplement sunlight and sun's heat.

Yield and Quality of Crops

Using tomatoes and cucumbers as an indicator for several kinds of vegetable crops, the plants will yield from 30 to 40 pounds apiece in a six- to eight-month season. With a population of 200 plants, this amounts to a total of 6,000 to 8,000 pounds of vegetables. It is possible to double-crop to some extent; for example, a crop of lettuce, radishes, or other quick-growing vegetables can be harvested before planting the main crop—or after the main crop is harvested.

The quality of hydroculture vegetables is superb. When properly grown, they are mineralized through both the soil and irrigation systems. Sustained nutrition with balanced plant foods enables the plants to form perfect cells in the vegetables. Flavor and "keeping" quality are exceptionally good.

The produce from the hydroculture units at Salinas, California, managed by Jim Wagner, is sold at premium prices in California vegetable markets.

Labor and Management Requirements

Excellent management is essential in a neighborhood hydroculture unit; less than this may result in loss of the crop and disillusionment of the neighborhood group of families. However, this enterprise lends itself to use of part-time duty in both management and labor. A skilled person can manage this vegetable-producing unit in only eight or ten hours a week. He or she can handle the operation with the assistance of two part-time persons. These might be people from within the neighborhood group.

Are there skilled people in local neighborhoods to handle a sophisticated enterprise like this? Surely they are available. American communities are filled with highly trained, underutilized people.

The labor needs in an operation of this kind are about five workers per acre during the planting and harvesting season. Using that as a guide, we estimate this small unit will require a total of 200 to 300 man-hours to produce and harvest the crops. After the area is planted, the operation will require about two hours of labor a day.

There are two obvious alternatives in meeting labor needs:

Free labor. If the families decide to provide the labor free, they may save 60 to 80 percent in costs of these foods. This might reduce the cost of tomatoes to fifteen or twenty cents a pound. However, this is a difficult plan to operate from the human psychological standpoint. It may be impossible except for close-knit fraternal or neighborhood groups.

Paid labor. If the families decide to pay for labor and management, they may save 30 to 50 percent of the supermarket cost of these vegetables. The savings are possible mainly because costs of marketing are eliminated.

Costs and the End Results

People considering local vegetable production should not delude themselves into thinking they may get lots of free vegetables. Cash must be paid for plants, seeds, fertilizers, utilities, rent, phone, supplies, and many other current expenses. Also, the building and equipment must be paid for —charged against the vegetables.

Management may falter; the families may quarrel over labor duty or sharing the vegetables. As a result, the tomatoes might cost a dollar a pound, not fifteen or twenty cents—or there may be no tomatoes at all, due to crop failure.

However, even with these hazards, the project may be quite worthwhile. The tomatoes will be gorgeous—red, juicy, and nutritious. Cucumbers will be a foot long, firm to the very tip and sweet in salads and sandwiches. Hydroculture is attractive because it can bring truly fresh vegetables into local neighborhoods again.

HOME GREENHOUSES

Home gardeners have three practical ways to extend the vegetable growing season, improving supplies of fresh produce even in winter months. They may use window boxes, which capture home heat; outdoor cold frames, which capture and hold the sun's heat in early spring and protect plants from harsh weather; and home greenhouses, which may be heated and lighted for vegetable growing all year round.

In a greenhouse the gardener can control heat, light, moisture, food, and air quality. For example, a heating unit set for a minimum of 50° will automatically provide that minimum temperature. The glass or plastic roof admits sunlight, but the supply can be supplemented with special electric lighting. In a sophisticated greenhouse project the operator may even govern the supplies of carbon dioxide for his plants; he may also use fans, ventilation, and misting devices to regulate humidity. These last controls, however, are a bit expensive, and they may use more energy than is advisable in a home food growing unit.

A hydroculture unit is a kind of greenhouse, but in most greenhouses plants are grown in raised beds and benches containing fertile soil rather than at ground level. Having the growing area waist high adds convenience to greenhouse gardening. The plants are nourished through garden procedures, not by means of irrigation water as in a hydroculture unit.

Regular potting soils from garden-supply markets are suitable for all types of home greenhouses. Some gardeners may wish to mix their own soils—especially for the larger size units—and the recipes and directions for making soils for minigardens in Chapter 6 may be used. The guides for making and using good fertilizers are also useful for greenhouse operations.

Small Window Greenhouses

There are modest little units often seen in home and garden magazines, where the window area is equipped to serve as a minigarden. Window greenhouses are usually glass

Window greenhouse

and are available in various sizes such as 33 by 52 inches and 48 by 60 inches, fitting many kinds of window areas. They include a soil box from 12 to 16 inches deep.

The window units cost from $200 to $250, exclusive of a heating unit, which may be provided for about $35. The total cost of operation may vary widely; greenhouse suppliers are willing to provide estimates for individual locations. Window greenhouses may be used to grow all kinds of salad vegetables, such as lettuce, radishes, green onions, zucchini, cucumbers, peppers, parsley, carrots, and tomatoes.

Lean-to Greenhouses

A lean-to greenhouse is a bit more ambitious, being built against an outside wall of a house or service building. Using the wall reduces cost and provides a measure of protection against cold weather.

A practical size for this kind of home greenhouse is 7 by 10 feet, providing 40 to 50 square feet of vegetable growing area and room to stand and work alongside it. Fertilizers and other gardening materials may be stored at one end of the work area. Smaller sizes are available, such as 4 by 8 feet, but they seem rather costly in terms of investment per square foot of production area.

A 7-by-10 lean-to greenhouse may cost from $750 to $1,000 for materials. The entire cost installed may run from $2,000 to $3,000. The smaller 4-by-8 unit costs about $600 for materials, or about $1,500 including electric services and installation costs. The actual costs of operation vary, depending on protection provided from outside weather by construction. Call a supplier for a specific estimate for your situation.

All kinds of salad vegetables may be grown in lean-to greenhouses. A typical unit contains a raised bench from 3 to 3½ feet high filled with a good greenhouse or potting soil. Many of the principles of minigardening are used to grow the vegetable crops.

Home-Sized Outdoor Greenhouses

These are typical greenhouses with raised benches like those readers have seen at nurseries and gardening places. Practical sizes for home use are 10 by 16 feet and 12 by 14

Lean-to greenhouse

feet, providing from 150 to 160 square feet of total area. About 50 or 60 feet of this is needed for work area, storage space, and a walkway, leaving about 100 square feet for growing vegetables.

These outdoor greenhouses cost about $1,400 for materials, and three times this for the complete unit installed and ready to operate—a total of about $4,200. These are only guide figures, of course. A greenhouse supply house can provide firm estimates of cost for your particular location.

Some greenhouse designs utilize milar or other plastic materials, rather than glass, for the roof and sidewalls. A plastic greenhouse is cheaper, due to less cost for framing and materials, and it may be quite durable. Greenhouse supply companies can provide information about glass versus plastic greenhouse designs and costs. There are also some greenhouses available that are constructed totally of plastic (many of those have frames that snap together). These are small and less expensive than the conventional greenhouse described above.

Excellent equipment is available to regulate temperature in outdoor greenhouses. The heating units with fans and ventilators may be set for a minimum temperature, such as 50° or 60°. A thermostat provides automatic control.

Operating such a heating unit to maintain temperature above 50° the year round requires about 300 gallons of fuel oil a year in a temperate climate zone such as New York or Kansas City. About 600 gallons may be needed to maintain the temperature above 60° in such a climate zone. Some greenhouse operators report that heating a 10-by-16-foot greenhouse costs about as much as for a two-bedroom house. This may be a bit high, but it illustrates that the heating costs are substantial.

Cooling by air conditioning or refrigeration is usually not required in temperate zones, such as New York, Ohio, Illinois, Colorado, and Oregon; however, it may be needed for summer operations in hotter areas such as Florida, Texas, and Southern California. Greenhouses are cooled with fans and ventilation in the temperate areas.

Home Greenhouse Yields

The yields of greenhouse vegetables are similar to those described under Hydroculture Units. Heated greenhouses may provide higher yields, since they may be operated for a longer season and double-cropped. However, the typical home greenhouse owner is after recreation as well as vegetables and he may not use soil as intensively as is the case with a hydroculture unit. Some of his little fields may be idle part of the time, or used for experiments and trying new methods and varieties.

Taking such factors into account, we suggest that the home greenhouse gardener expect to produce about 5 pounds of vegetables per square foot of growing area in a season. Then, if the average price of salad vegetables in food markets is forty cents a pound, the home crop might be worth $2 per square foot. The yields from the three kinds of home greenhouses might be as follows:

UNIT	POUNDS OF SALAD VEGETABLES PER SEASON	VALUE AT RETAIL PRICES
A window unit with 12 square feet of planted area	60	$ 24
A lean-to unit with 50 square feet of planted area	250	$100
A conventional home greenhouse with 100 square feet of planted area	500	$200

As always, these are only guide figures, attainable but far beyond the performance capabilities of many home greenhouses.

The precious dividends from home production of vegetables in greenhouses are the same as from all kinds of home gardening: the money value is only a part of the benefits. The balance is provided in freshness and flavor—a sweeter cucumber and a fine tasting red tomato, and in vegetables that provide you with full cargoes of health-giving minerals and vitamins during many months of the year.

This Certifies that
WENDY KERLIN
planted and harvested
67 lbs. of vegetables
in Merkin, Connecticut
in 1975.
Harvey T Grossman
COMMISSIONER

9.

YOUNG PEOPLE'S GARDENING AND FOOD PROJECTS

In 1968 one of the authors was living in Washington, D. C., in its trouble zone where massive riots had leveled three major streets and produced over $100 million in damages and property losses. One cause of riots was a tragic movement of thousands of farm and rural families into the heart of Washington, although they had no places to live or ways to make a living.

During this troubled time a fourteen-year-old named Allen came to see the author. At first the young man only sat and talked, and it was difficult to know why he had come. Then he brought a surprisingly good picture he had drawn and offered it as a gift.

After these amenities Allen said, "I hear you know about farming and gardening and I want to learn to be a gardener. Will you help me?"

"Yes," the author agreed, "I'll help you, but why are you interested? Do you have a garden? Do you come from a farming family?"

He said, "No, that's the reason I need help. I have never been in the country and I really don't know how things grow. But I want to learn. I have a garden in our front yard, but my father hates it. He says farming enslaved our people, but I don't believe that. Anyway, will you come and see my garden?"

The author did, and the end result two months later was a visit with this young city gardener to the director of Washington's Botanical Gardens, where they hired Allen to work during summer vacation in the plant-propagation and nursery center.

Allen was not a freak. Millions of America's urban children want to be gardeners—if only they all could have a chance.

We will discuss young people's gardens in this chapter, using as examples Washington's program, Cleveland's school-based project that includes over 20,000 young people, and Eddie Albert's Pocket Farms for Young People. Many more good projects might be described, but these will illustrate youth gardening in America, one of our country's fastest growing movements.

WASHINGTON YOUTH GARDENS

Washington Youth Gardens was established in 1962 by Mrs. Martin Vogel. Impressed by a similar program in New York, she provided the spark of creation and personal interest that is so necessary in starting useful enterprises of this kind. Mrs. Vogel also put the Youth Gardens on a sound legal and business basis by incorporating it under nonprofit laws. Thereafter, this organization was able to serve as a competent sponsor of young people's gardening projects, and to receive and administer private contributed funds to assist in their success.

During twelve years the Youth Gardens have attracted increasing support, so they now have more than seventy officers, sponsors, and advisers, including several members of Congress. Mr. Robert F. Lederer, head of the American Association of Nurserymen, serves as Vice-Chairman of Washington Youth Gardens. Young people's gardening deserves this kind of blue-chip support in any city where it is conducted.

Ground Rules—How the Gardens Operate

William (Bill) Hash is the Director of the Washington Youth Gardens projects; Jerry Smith is Assistant Director. They launch the operation each year with an enrollment of about 1,000 young people from District of Columbia public schools and playgrounds. Six inner-city schools provide 300 of these young gardeners. Bill Hash and Jerry Smith visit the third, fourth, and fifth grades in February and early March to present the program and enroll these young people. In the meantime directors of City Park and Recreation units use similar procedures to enroll about 700 additional students in an age group of about eight to fourteen years.

To describe the rules and procedures, we will present a typical discussion during a school visit.

Garden Areas

The Washington program is conducted on three main sites: a two-block area near 14th and Taylor Streets, called Twin Oaks, in Washington's inner city; a land unit at the National Arboretum, in the nearby suburbs; a property in the Anacostia area of the inner city. These lands comprise about five acres and are provided without cost since they are public properties. A study of the District of Columbia in 1968 showed that ample idle and underused land was available for youth gardening projects, if approving bodies would act favorably to provide it. There has been no shortage of land for the Washington Youth Gardens program. Transportation in school buses is provided for children who live too far from designated sites to walk to them.

Preparing the Land

Each spring under Bill Hash's direction, the city plows and rototills the garden lands. Lime and fertilizer are applied at this time, too; then the areas are subdivided into about 1,000 individual plots, marked, and numbered. In addition, some suitable land is prepared for several larger plantings of certain vegetables to be grown and harvested by groups of gardeners. These larger plots are usually planted to cucumbers, corn, and summer and winter squash.

Planning and Orientation

Ten years of experience with Washington children shows that you don't plan abstractly but in a very concrete way. The work commences with groups of the young persons learning how to measure and stake out a plot of soil. A typical plot is about 5 by 14 feet. Then the advisers work with each gardener in planning his or her own plot.

AN ENROLLMENT SESSION AT A SCHOOL

Teacher: We have Mr. Hash and Mr. Smith here today to tell us about the Youth Gardens. Some of you may like to join this year and grow gardens. If so, they may be able to enroll you. They will tell you about it.

Bill Hash: Some of you know about Youth Gardens since you were in it last year. I have some slides here that show some of the gardens and what we did.

(He shows color slides and explains them for ten minutes.)

Bill: Now you get the idea, you join and get a garden plot assigned to you. We plan it together. You plant the seeds and take care of the plants. You grow vegetables—corn, collards, cabbage, string beans, summer squash, onions, beets, lettuce, radishes —lots of things.

Student: Where do we get the seeds?

Bill: We provide them. And we dig up the ground with a tractor and fertilize the soil. Then we stake out the plots that you will use.

Another Student: Can I choose my plot?

Bill: Well, we have lots of children and don't want any quarrels, but we'll try to have a good one for you.

Student: Isn't it lots of work?

Jerry Smith: Sure. But it is good work. You grow things.

Student: Do we just take the vegetables home?

Jerry: Yes. And we have some scales to weigh them and a produce chart. Every time you pick some vegetables we weigh them and give you credit.

Bill: As I said, each student that joins this program tries to grow at least forty pounds of vegetables. Some grow much more. Last year one of the gardeners grew 110 pounds and got first prize.

Student: How do we decide what to grow?

Jerry: We help you to plan your garden. But we also find out what your parents like and don't like for you to bring home. It's shown on your permission sheet.

Student: I don't like beets.

Bill: Then don't grow them. Grow string beans. There are lots of other vegetables.

Teacher: Now I hope you have enough information to tell your parents about this, and I hope we will have quite a few gardeners from here. But everyone that joins will have to have permission from their parents. And here are the permission sheets.

Bill: How many of you think you would like to have gardens this year—hold up your hands. Then you can take home a permission sheet for your folks to sign, if they want you in this program.

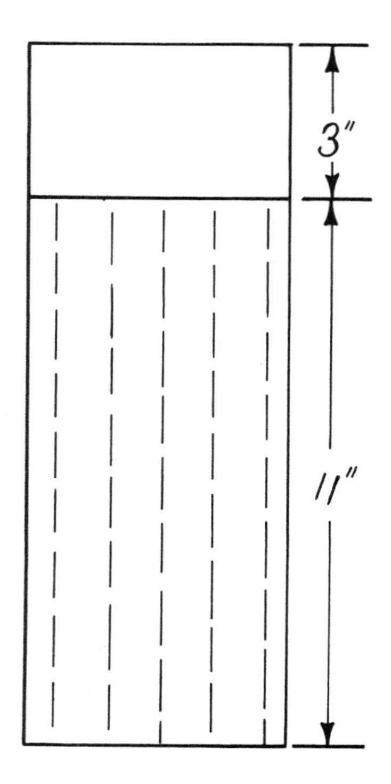

Individual garden plan

Planting and Growing the Vegetables

About 100 schoolchildren have enrolled in Twin Oaks Youth Gardens. Bill Hash and Jerry Smith have divided the gardeners into two groups so individual attention can be given each youngster. Three teenage part-time assistants are employed to help with the operations, so there are about ten children per supervisor.

Prior to planting days the garden seeds are bought in bulk and measured out into marked envelopes for the children by the assistants. On a typical planting day in April groups of noisy, eager, mischievous children (an average group has eight- to twelve-year-olds) gather at Twin Oaks. Seeds and tools are distributed, and the children go to their numbered plots with their supervisors.

First, the child and supervisor measure a portion 3 feet from the end to reserve for tomatoes, cabbage, peppers, and eggplants, which will be transplanted later. Then they mark the little rows 1 foot apart and 10 to 11 feet long for other vegetables to be grown from seed. The April plantings may include lettuce, radishes, beets, carrots, green onions, collards, chard, mustard, or spinach—whatever kind of greens the child's family prefers. Beans, corn, cucumbers, and okra come later.

In May, June, and July all the young gardeners cultivate and weed their gardens and harvest the early kinds of vegetables.

Harvests and Awards

Getting a good yield is a major aim of every child in the Washington Youth Gardens program. Each young gardener tries to harvest at least 40 pounds of vegetables to take home; some achieve more than twice this amount.

The crops are checked and weighed on harvest days and marked on everyone's produce chart. At season's end totals are made for each child. Every gardener receives a signed certificate of achievement, and there are blue ribbons and trophies for the gardeners at each center who grow the most vegetables, and who grow the finest of each kind.

In 1973 the largest harvest by a young gardener was 110 pounds, taken home to improve family food supplies and save on food costs. At the present-day average value of thirty cents a pound for fresh vegetables, this amounts to $33. A usual child's achievement of 40 pounds is worth $12, or twice that if you count improved nutrition of the family. Of still greater value is the young gardener's personal success, important these days for many children in urban society.

The Fall and Winter Projects

It would be wasteful, indeed, to build this good relationship with young people in springtime and let it evaporate in fall. So Washington Youth Gardens simply changes with the seasons. In October attention shifts to making jack-o'-lanterns, dish gardens, terrariums, and holiday items.

Bill Hash and his coworkers go to the schools again to recruit this group, many of whom grew gardens in the spring. They enroll about 200 children in the fall and winter projects. In addition to the activities listed above, some of those in the group seriously interested in gardening learn about seed testing, plant propagation, and soil analysis. Others learn how to plant bulbs for early spring blooming, and how to make good flower arrangements.

Slides and movies are used extensively in the fall and winter phase, and there is a need for video materials suited for young people in these useful projects.

The annual budget of the Washington Youth Gardens program is about $50,000, used mainly for salaries. The operations are underfinanced. Available funds should be increased by about 50 percent through tax-exempt grants and contributions. We think this is a good civic investment in Washington, D. C.

CLEVELAND'S YOUTH GARDENS

Established in 1902, the Cleveland Youth Gardens program is over seventy years old, the oldest and, in some respects, the largest and finest in America. Over 20,000 children and young people are involved. During 1974 these young gardeners grew and harvested over $500,000 worth of vegetables.

Paul Young, gardener emeritus of Cleveland, is a distinguished leader in this development. He introduced the "tract garden" idea around 1925, has supported youth gardens all these years in news and public media, and has helped Cleveland Youth Gardens to achieve a high professional status.

Many years ago these youth gardens were reinforced by being officially based in the public school system. Garden science was added to the science curriculum of the fourth, fifth, and sixth grades. Vocational horticulture was added to high school studies, and adult training also became available in this city's educational system.

An interesting feature of the garden science courses is closed-system radio broadcasting into schoolrooms to assist in teaching how to prepare soils, plant seeds, and care for successful gardens. Through such devices a fund of knowledge and skills is built for use by the young gardeners in springtime.

Each year the young people in Cleveland's gardening program are invited to exhibit their finest produce. The garden exhibits are widely publicized, and there are awards and ribbons given to the finest gardeners. Every young person who completes his gardening project for the year receives a Garden Certificate issued by the horticulture department of the Cleveland Public Schools.

Cleveland's Youth Gardening is divided into three main phases:

Tract Gardens

Launched in the 1920s, this phase is conducted in fourteen garden areas in different parts of Cleveland, comprising about 32 acres. All ages of schoolchildren participate, with the youngest cultivating 3-by-6-foot sizes and the oldest handling tracts as large as 12 by 30, about as large as the model home garden in Chapter 2. About 5,000 children grew tract gardens in 1974.

Home Gardens

These are young people's home gardens, made successful by skills taught in the public schools, supplemented by parents' and neighbors' assistance. In addition, over 100 teachers visit the home gardens in late spring and summer to give technical assistance and encouragement. Three visits in the gardening season are desirable to promote full success. Due to budget limitations and a heavy enrollment, it is possible to make only one or two visits to assist these young gardeners.

About 11,300 young Clevelanders participated in this phase with full-season gardens in 1974; about 6,500 children grew short-season minigardens, or salad gardens. The entire home garden phase is popular, a confirmed success in hundreds of Cleveland's neighborhoods.

Vocational Gardens

These are gardens planted and managed by young people in training for employment in floral, nursery, landscape, and environmental fields. They are of high school and post-high school ages. This phase included 606 young persons in 1974.

Cleveland's Example

Cleveland's Youth Gardens have the marks of success; features that seem worth adapting and using in many cities and towns of America. Most significant, the program has strong roots and resources in the municipal school system.

Being substantial and mature, this program has able leadership illustrated by Peter Wotowiec, the director. He says, "Cleveland Youth Gardens included about 24,000 children and young people this year (1974). The values to them and their families are, of course, unmeasurable, since they include contacts with nature and acquisition of skills that are useful lifelong. Children who learn gardening are well on their way to becoming good citizens.

"Many people ask me what this gardening program is worth—how much food it produces," continues Peter Wotowiec. "I tell them that the money value is the lesser one—that building good children is more important. However, there is little doubt that these young people produced and helped to consume over $500,000 worth of vegetables in 1974. This helps a great deal in home foods. The children in this program are typical of the whole city. They come from modest homes that need these foods for nutrition as well as to help home food budgets. Let's just say this is a million-dollar project. That means the average child in it created $50 for his parents and the city of Cleveland.

EDDIE ALBERT AND POCKET FARMS FOR YOUNG PEOPLE

Who is better situated than Eddie Albert to help young people become "pocket farmers"—growers of vegetables for home use? All of America knows Eddie as the gentleman farmer of *Green Acres* on TV.

Four years ago Eddie went to Champaign, Illinois, to help a group of young people set up a gardening project using land in the city parks. "My role," he said, "is to help in publicizing this project so these young people can get resources and encouragement they need to be successful."

He did his work. The Champaign project has expanded from about seventy-five plots in 1971 to over 300 in 1974, with whole families joining the children in vegetable gardening. Awards are given, and the young gardeners wear Eddie Albert T-shirts.

From this beginning Eddie is currently developing one of America's most interesting youth and food programs. It is based in his home place—in the "greenest acre"—near Santa Monica, California. He uses the most up-to-date techniques in ecological gardening, including a demonstration hydroculture unit for growing highest quality tomatoes and cucumbers. Film strips are made for use in promoting successful youth gardens in many cities and to help in obtaining essential resources. San Francisco's "pocket farms" program for young people aged seven to seventeen is only one example of a growing project.

Currently Eddie Albert is the most active and successful promoter of youth gardening in America.

These youth gardening programs illustrate what is going on in dozens of major cities from Boston to Cincinnati to Los Angeles, including hundreds of rural and suburban neighborhoods in between. It is a beautiful trend that should be encouraged for sociological reasons as well as for coping with home food problems.

TOMATOES
PEAS
STRING
BEANS

10.

PROCESSING AND STORING THE HARVEST

If your crop fails, you have nothing, but if the plants yield well, you have a new problem: what to do with all those vegetables. Processing and storage is the answer. People have been storing foods against wintertime needs ever since the beginnings of civilization. We share this trait with squirrels, bees, ants, gophers, and dogs that store food for later eating.

It is only in recent years that Americans have laid aside their individual food-storage skills. The amazing yields of good home gardening demand that we use them again. In the present chapter we will identify the various ways to store vegetable crops and provide a number of useful guides and recipes.

CANNING FOODS IS BASICALLY EASY

The canning of foods, originated in 1810 by Nicholas Appert, involves only two actions: killing bacteria with heat and removing excess air from the cans or jars.

As many of you will remember, your mothers or grandmothers canned fruit by simply boiling it, pouring it into glass jars, and sealing the jars. This served to kill the bacteria and exclude air.

In the case of vegetables a bit more is involved. Since yellow and green vegetables, unlike fruits, are nonacid foods, botulism spores, which originate in the soil, have a better chance for survival. Vegetables canned carelessly or with

FOOD STORAGE THROUGH HISTORY

The ancient Egyptians who put grain in Pharaoh's tomb for him to eat in the hereafter were only illustrating the food arts of their time. They stored grain and other foods as early as 5,000 B.C. and commenced making beer from sprouted barley in 3,000 B.C.

The Chinese used soybeans more than grain as a staple food. By 2,000 B.C. they had learned to use these nutritious legumes in making soy sauce and soybean curd, storing the curd in cool dark places for winter's food.

American Indians used food storage arts on meats, dried berries, and camas roots. Storage of vegetables was often limited to maize and beans, since these were easily dried and kept in pots or pouches. They served as staple foods among tribes of North and South American Indians and were introduced into Europe from here. *History of Foods* by Reay Tannahill (New York: Stein & Day, 1973) reports that in 1543 Columbus carried vegetable seeds, chickpeas, wheat, and sugarcane back to Europe from North America. Later maize, beans, potatoes, peanuts, peppers, tapioca, and manioc were sent over as new and storable foods for European diets.

Fall in Oregon in 1920 was many food years ago, but our memories are sharp and clear. The family food habits came from Kentucky in 1852, when grandfather moved west and settled in the Willamette Valley. Seventy years later his arts in handling and storing foods were used in our family.

We had three gardens: Mother's salad garden by the back porch; a family garden nearby; and a big garden for winter vegetables on the fertile riverbank half a mile away. Canning and pickling season commenced in June with early fruits, and by October our big basement storage area was stacked with row on row of beautifully packed fruits, vegetables, pickles, relishes, and preserves. And there were sacks of potatoes and pounds of dried vegetables and fruits. We knew we could find our way through winter without serious danger of starvation.

How times have changed in America! However, we seem again to be on an upswing in improving common skills of living, eating, and working. Millions of young people are learning to grow foods, and they will also learn the arts of storing them for winter eating.

inadequate heat may act as an incubation site for the botulism spores that grow into poisonous bacteria. If good equipment is used and precautions are taken, however, home canning of vegetables is quite safe. The heat required to kill these spores is 240°F. for ten minutes—twenty minutes for corn and spinach. This requires *pressure* cooking—just boiling at 212° for a long time is not enough.

Pressure cooking is basically easy, too. Millions of American homekeepers learned to can vegetables with pressure cookers during the Depression years of 1930–1940 and in World War II, when food was scarce. They learned to do it well and to avoid dangers of food poisoning.

Home canning with a pressure cooker requires (1) a 16- or 21-quart-size pressure cooker, costing from $40 to $60; and (2) glass jars costing about $3 a dozen for quart size and $2.25 for pint size,

Canning String Beans

1. Wash and check the glass jars and lids.

2. Wash, snap, and prepare the beans; drain them.

3. Pack the beans snugly into jars, but leave about 1 inch of head space. Add 1 teaspoon of salt per quart jar. Cover the beans with boiling water, and let any air bubbles escape.

4. Put on the jar lids and screw down firmly.

5. Pour hot water into the pressure cooker, following directions of the manufacturer—usually 2 to 3 inches in bottom. Lower the filled jars into the cooker. A 16-quart cooker holds 7 quart jars, or 7 pints. A 21-quart cooker holds 7 quart jars or 18 pints (being taller, it holds pint sizes in two layers).

6. Following the manufacturer's directions, cook the jars for ten minutes at 240°F. (10 pounds of pressure on the gauge). Corn and spinach must be cooked for twenty minutes at this temperature.

7. Release pressure and remove lid, following directions. Remove jars. Check next day to see that all jars are properly sealed, and store them in a cool, safe place.

Glass jars may be reused indefinitely, as long as they are not chipped or cracked. New lids are used at a cost of about $1 a dozen at the next canning season.

Information from a glass jar maker shows that home-canned string beans cost less than half the price of grocery store beans, even when the fresh beans are purchased. If you raise your own garden beans, the cost of a year's supply of canned string beans for a family of four is only about $7, one-sixth the cost of the same amount in a food market.

References

Many useful bulletins, books, and leaflets on home canning are available, including those provided by manufacturers of canning and freezing equipment. These will give you a start:

"Home Canning of Fruits and Vegetables." USDA Bulletin. Write to U. S. Department of Agriculture, Washington, D. C. 20250.

Complete Cookbook of Home Storage of Vegetables and Fruits, by Evelyn Loveday. Charlotte, Vermont: Garden Way Publishing Co., 05445. $3.00.

The "Blue Book." Ball Corporation, supplier of home canning equipment and supplies. This is mainly on home canning. Write to Ball Corporation, Box 2005, Muncie, Indiana, 47302. $1.00.

"Kerr Home Canning and Freezing Book." Consumer Products Division, Kerr Glass Mfg. Co., Sand Springs, Oklahoma, 74063. $1.00.

Farm Journal Cookbook of Canning and Freezing. New York: Doubleday. $5.95.

A NEIGHBORHOOD CANNING PROJECT

This kind of a local enterprise became quite popular in 1940–1950, due to the stimulus of World War II and its food shortages. Once started, some of these neighborhood canneries persisted into the 1960s. With a renaissance in home gardening, neighborhood canning may again be of interest.

The Group

It could be a neighborhood gardening group of 100 families or more; or a fraternal, church, or other fairly stable group; or gardeners of an area who wish to patronize a custom cannery. The main idea is to save food money, since people who have low-cost vegetables from their gardens will want a canning or freezing service in order to cash in the year round on these low-cost food supplies.

The Facilities

A neighborhood cannery is made up of the following:

A work and cannery space or building
Worktables for use in preparing the foods to be canned
A steam tunnel to preheat the filled cans and expel air from them
A semiautomatic can sealer
A steam cooking retort, which is a large pressure cooker
Baskets to hold the cans
A mechanic's hoist to lift the baskets of cans in and out of the cooking retort
A small steam boiler to provide steam for the retort and steam tunnel and for cleaning the cannery.

As small factories go, this is quite a simple, low-cost kind—obtainable if really needed.

The Operating Plan

The neighborhood cannery has a trained manager and assistants, if needed.

The local people make appointments to use the cannery. They take their produce—peas, beans, cherries, corn, or whatever—to the cannery and prepare it, filling their own cans and marking them in indelible ink.

The manager processes, seals, and pressure cooks all of the foods, using good cannery practices and controls.

The users come back and pick up their own canned goods when ready. They pay a fee per can; for example, fifteen cents a can for string beans for the cannery service.

Such a local custom cannery might be a small co-operative or a private business owned by a local person. If so, he might do other canning under commercial contracts.

A COMMUNITY CANNERY

The Discovery of Neighborhood Talent

We live in a country of richly developed people, millions of whom have learned the skills of industry and commerce, but have diminishing opportunities to practice these skills in serving employers or their communities.

We learned about local supplies of underutilized people when we built Garden Home Community Cannery in 1942, near Portland, Oregon. Food was getting scarce due to the war and home production was the "in" thing. We had a neighborhood meeting and planned the local cannery.

Needing capital, we launched a campaign to sell co-op stock at $10 a share. Zealously, with huffs, puffs, and arm-twisting, we amateurs sold fifty shares and had a measly $500 toward a $5,000 budget. Then a neighbor named Wittenberg came along and sold 450 shares in two weeks, handing over $4,500 in cash for the project. He was a former bill collector for the gas company who knew how to separate people from their money.

We had to plan the cannery building, a cinder-block structure with a trussed roof. An engineering company said they'd have the plans and specifications ready in sixty days. Meanwhile, a casual neighbor named Mac came along and built the thing, finishing it the same week we got the plans from the engineering company. It turned out he used to design and build roofs for a big company in the Midwest.

With the building ready, it was time to install plumbing. Our brave crew got pipes, tools, cutters, nipples, and plumbed half of it in ten days. Then a quiet neighbor came along and said, "Want me to help tomorrow?"

Next day he came with an acetylene plumbing rig and finished the whole job in two days. He used to be a gas company plumber, doing high-speed work on live gas lines.

Well . . .

Came time to start the cannery, which we did with self-trained volunteer help. Then a neighbor came along who proved to be a senior canning technologist, a connoisseur of canning who hated big canneries and loved small ones. Thereafter, he ran the Garden Community Cannery and trained all the skilled persons we needed.

This is the story of America; there are plenty of skilled people to build and run canneries, hydroculture units, and food enterprises in neighborhoods all over the country, *if* the vegetables are really wanted and needed.

The problem is not in a supply of skilled people; it is in finding ones who can get along with their neighbors where work, food, and money are concerned. Dividing up a crop of vegetables is, after all, a supreme test of skill. If we can learn to do this, the rest is easy.

FREEZING GARDEN VEGETABLES

Freezing garden vegetables at home is even easier than home canning. However, the cash costs of owning and operating a good home freezer are rather high. Also, there is a risk of electric power failure to be considered—possibly an increasing risk due to national and local electric energy problems. Home freezing of vegetable supplies is suggested only if you can afford the investment and feel reasonably sure there will not be a power failure in your area.

Freezing Procedures

1. Wash and prepare the vegetables as for home cooking. The home freezer handbook provided with your freezer gives instructions.

2. Blanch the vegetables by immersing them in boiling water for about three minutes, using a wire mesh basket or strainer. Again, follow instructions in freezer handbook or other guide for specific kinds of vegetables.

3. Cool the blanched vegetables in cold water, then put them in plastic freezer bags in convenient family meal quantities; for example, about 1 pound per bag. Tie or close the bags with rubber bands or "twist-ems." The packages may be put in freezer boxes, if desired, or stored without boxes. If they're in boxes, label them.

4. Put the packages or boxes in the freezer for safe storage until use.

Home Freezers and Costs

The freezer is the only major equipment item needed for home vegetable freezing. A popular size is 15 to 16 cubic feet, holding an estimated 500 pounds of foods. At present-day prices the cost of such a freezer is about $275 to $300—about $30 a year if you expect to use the freezer for ten years.

The operating costs for a 16-cubic-foot home freezer will vary with local costs of electric power; however, manufacturers' information and USDA guides indicate that the operating costs per year may be approximately as follows:

Electricity	$40.00
Financing	20.00
Repairs and miscellaneous	6.00
Add annual depreciation	$30.00
Total	$96.00

A home freezer holds meats, fruits, and other foods besides vegetables. Also, a family fully using its freezer may fill it and use up the foods more than once a year, thereby reducing the freezer costs per pound of stored food. However, the usual costs per pound for storing frozen foods may be approximately as follows:

Amount stored in 1 year	500 lbs.
Annual costs of owning and operating a 16-cubic-foot freezer	$96.00
Cost per pound	19.2¢

If you do not use your freezer efficiently—for example, storing only 250 pounds of various foods in a year, rather than 500 pounds—the storage costs may rise to forty or fifty cents a pound.

These rather high costs for storing frozen foods cause many families to count the home freezer as a wonderful convenience, but not a cost saving in the home budget. The exception may be if you have access to meat at farm wholesale prices, in which case the savings can help to pay the costs of storing fruits and vegetables. Also, if you have regular access to low-cost fruits and vegetables in season, and have a large household of people to feed, your savings will be substantial.

In these favorable situations a combination of canning and freezing may serve in providing supplies of good foods at low costs. More information on freezing is available in "Home Freezing of Fruits and Vegetables," Home & Garden Bulletin No. 10, U. S. Department of Agriculture, Washington, D. C. 20250 (35¢).

DRYING HOME FRUITS AND VEGETABLES

Many of our mothers and grandmothers did this, and we can still remember the exquisite flavor of sun-dried black raspberries and sweet corn. They also dried apples, pears, peaches, and prunes.

How to Dry Corn

Present-day home gardeners have easy access to corn. For their possible interest in drying sweet corn, here is a working model.

1. Gather the corn, husk and silk it.

2. Using a sharp knife, cut the corn from the cobs into any convenient pan or kettle.

3. (old-fashioned way). Spread a sheet of cheesecloth or other porous cloth on a wire mesh tray; then spread the corn in a thin layer on the cloth. Put out in the sun daily until the corn feels quite dry. This way is useful only if you have fairly clean air and a plentiful supply of sunshine.

3. (second best way). Spread the corn on the bottom of a large shallow pan or of several pans or cookie sheets. Cover with cheesecloth for protection against soot and dirt, and put in a warm place or in the hot sun. Stir the kernels occasionally. Within a week or so, the corn will become dry, though a bit sticky due to presence of corn sugar.

4. Store the dried corn in airtight jars or cans until time to use. Soaked a bit before cooking, it makes a delicious vegetable food.

Drying and Storing Beans and Peas

This is even easier than drying corn. First you should study seed catalogs and select the kinds you want; for example, great northern white beans, limas, red kidney beans, or chickpeas. Obtain seeds and plant a little field of them in your garden, following seed packet directions. Instead of harvesting them green and tender, allow them to mature fully until the pods get dry and crackly. On a warm sunny day the crop may be harvested by picking the pods into a kettle or sack.

Separate the beans or peas from their pods, put them in a big cloth or plastic bag, and roll them on the floor, pressing the mass so as to break the pods and liberate the seeds. Gently walking on them may help to do this.

Then, in a windy place (or using a fan), pour the beans or peas, crushed pods and all, from about 6 feet high onto a clean cloth or paper. The wind will blow the pods and litter to one side, while the beans or peas will drop nearly straight down, being heavier. Thus, with repeated operations, you may thresh your crop, just as the Indians did 800 years ago.

If the seeds seem at all soft—if they can be dented with a thumbnail—dry them some more in a warm, ventilated place. Then store them in sealed jars to keep weevils from eating them.

PICKLING

Pickling can lengthen the life of your crops and add interest to your meals. Cucumbers, cauliflower, peppers, onions, cabbage, beets, tomatoes, and other garden vegetables can be pickled and stored easily by the home gardener. Since cucumbers are the most popular of the pickling vegetables, we will focus on their use in the home. Consider the following assortment, and possible cost savings:

	Retail Value
10 qts. dill pickles	$ 9.70
20 qts. mixed sweet pickles	24.00
15 pts. hamburger relish	12.00
10 pts. bread and butter pickles	5.70
(or your choices)	
Total	$51.50

Such a delectable pickle supply, made of your surplus cucumber crop, can be provided at a cash cost of less than eight dollars.

Pickling is basically a process of fermentation caused by bacteria. Salt or vinegar is used to control the fermentation process. Thus, we have salt brine pickles, vinegar pickles, and combinations of these. The addition of sugar, spices, and dill provides desired flavors.

Genuine dill pickles are made with a weak salt brine and a bit of vinegar. These mild agents permit a good fermentation of the cucumbers, giving the old-fashioned dill pickle texture and flavor.

Traditionally, home pickles are made in heavy crocks of 5-, 10-, and 20-gallon sizes. However, these are expensive and hard to find. If unavailable, you may use plastic trash cans. They are clean and chemically stable, quite satisfactory for pickle making.

Basic Pickle-Making Guide

Here are the basic steps in pickle making, using the salt brine method. Start with a mild brine, increasing its strength as you add more cucumbers.

1. In a separate container make a mild salt brine by dissolving 5 pounds of salt in 5 gallons of water.

2. Trim the dried blossoms from the first picking of cucumbers (they harbor undesirable molds and enzymes), and trim the stems. Wash thoroughly.

3. Clean your plastic can, and put the first load of cukes in the bottom. Add brine to cover them. Also add a quarter cup of dry salt per quart of cucumbers to increase the strength of the brine.

4. Cover these cukes with a sheet of plastic to serve as a water seal, excluding air. This is necessary to prevent mold from forming on top of the brine and cucumbers. To do this cut a piece of plastic twice as wide as the diameter of the can. Press it down so that it touches the brine, and add a gallon of water. The weight of the water on the plastic sheet holds it down and against the sides, excluding air.

5. At the next picking of cucumbers repeat the above actions: trim and wash them, remove the plastic water seal, add the cukes, add brine to cover them, and add a quarter-cup of salt per quart of new cucumbers added. Then replace the water seal.

6. Repeat the above until your cucumber crop runs out or your container is full. Cauliflower, onions, small carrots, peppers, and other vegetables may be brined along with the cukes, if desired. Be sure to keep the cucumbers and other vegetables submerged in brine at all times.

7. Let the pickles ferment and ripen for seven weeks after the last addition of cukes and salt. They are then ready to use as plain salt brine pickles.

Once you understand the basic process, you may add vinegar, spices, and various vegetables to suit family tastes.

Dill Pickles

To make good old-fashioned dills, follow these guides:

1. Run 10 gallons of water into a clean plastic can. Add 7½ cups of dry salt and stir to dissolve thoroughly. (Or put the salt in a cloth bag and hang in the water until dissolved.)

2. Add and mix thoroughly into the brine: 5 cups white vinegar; 6 or 8 dill heads and crushed dill stems (if unavailable, add dill flakes or powder from a food market); 5 or 10 cloves of crushed garlic, if you like it; 2 tablespoons mustard; 2 tablespoons celery seeds.

3. Trim and wash cucumbers to fill the pickling can. Sizes above 3 inches long are preferred.

4. Install a plastic water seal, as in making salt brine pickles, to be sure the cucumbers are submerged at all times.

5. Let the pickles ferment and age for four to six weeks before starting to use them. They may be stored in a cool, quiet place to improve and be eaten while life goes on.

These basic recipes will open the subject of pickle making, illustrating that valuable home foods can be made from surplus garden vegetables. Don't be too kind-hearted and give away all these vegetables to friends and neighbors. Store plenty for your own family; it may be a hard winter.

Readers who wish good recipes for sweet, sour, bread and butter, and other pickles and relish may write to the authors at the address given in Appendix A.

SAUERKRAUT

Sauerkraut is made by a mild fermentation of cabbage, using salt to control the process, just as pickles are made by mildly fermenting cucumbers.

Procedure

1. Wash and prepare a plastic trash can or an earthen crock. Choose a size suitable for the task. A 10-gallon size may be sensible.

2. Gather firm heads of fall cabbage. Trim off outer green leaves, which yield dark sauerkraut, and store the cabbage heads in a ventilated dark area at room temperature for a week. This prepares the cabbage for easy cutting into shreds.

3. Cut the cabbage into shreds with a sharp knife or a cabbage cutter. Make fine shreds not over 1/16 inch thick. Do not include the cabbage cores.

4. Salt the cabbage shreds. Accuracy is necessary, so use scales and a 3- or 4-quart mixing pan. Weigh out exactly 2¾ pounds of cabbage shreds and put in the pan. Add exactly 1 ounce of salt, and mix until every shred of cabbage has particles of salt on it.

5. Pack the salted cabbage into the plastic can. As the 2¾-pound units are added, stomp them down with a big wooden potato masher, a baseball bat, or other useful wooden weapon. This starts the cabbage juices running and eliminates air, promoting good sauerkraut.

6. Keep adding and stomping until the can is nearly full. If natural brine in the cabbage does not cover the mass, add some. Make it with 3 ounces of salt per gallon of water, and add enough to cover the cabbage.

7. Install a plastic water seal, as described under pickle-making, and store the kraut-to-be in a cool quiet place of 55 to 65 degrees while nature does its work. By magic, you will have delicious sauerkraut in four to six weeks, and it improves with age.

Sauerkraut may be used from the original container as long as it is well stored and carefully sealed to prevent exposure to air. It may also be put in tightly packed jars and kept in a refrigerator, or, it may be sterilized and kept in jars. To do this, heat to at least 165 degrees in a kettle (use a roasting thermometer) and pack in quart or pint jars. Seal with sterilized lids.

STORING VEGETABLES IN A CLAMP

Clamp is the Old English name for a place where you store potatoes, turnips, rutabagas, parsnips, cabbage, pumpkins, and squash safe against cold winter blasts.

It is a hideaway for rough, solid garden foods, much like the storehouse a squirrel makes for nuts or a mouse for grains of corn he stole from the farmer.

Traditionally a storage clamp is made in the garden or a place with soft soil. Dig a hole 2 or 3 feet deep and line it with 6 inches of peat moss, dry shavings, or straw. Then pack clean vegetables in the hole and cover them with more peat, shavings, or straw and a layer of burlap or cloth; finally, add 8 to 10 inches of soil.

These materials insulate the foods from harsh weather. Occasionally you may open the clamp and remove a supply, then carefully restore the covering. Root vegetables may be kept for three to five months in this way.

Another method of storage is to put potatoes, carrots, rutabagas, and parsnips in a basement place where they will be cool but safe from frost. Pack them in straw or peat and sort occasionally to remove spoiled goods and sprouts from the potatoes.

To prolong the storage life of squashes, sterilize them by washing lightly with dilute formaldehyde to kill mold spores. Store in a cool place on shelves.

THE REWARDS OF FOOD STORAGE

Many Americans seem unaware of the fact that a great deal of the eating is unnecessary. Fat people are eating too much, and hungry people are eating many useless foods. As prosperity winds down and inflation winds up, the efficiency of our American system of life will be tested by our ability to move over onto better, simpler foods.

Part of the test is in what we store as families and living groups for winter eating. Can you imagine an intelligent person storing TV dinners, pot pies, white bread, and plenty of Coca-Cola? No, storage is for simpler basic foods.

Dick Gregory has settled on dry beans as a staple food in his storehouse; he has stashed away 2,500 pounds of various kinds to carry his family and friends through bad times. It is a good choice, since beans provide protein, calcium, phosphorus, iron, vitamin B, and many trace minerals. Along with a bit of egg, fish, or chicken and home-grown greens, you can live quite well on beans for a month, a year, or even a lifetime. And did you know that rutabagas have calcium, phosphorus, iron, vitamins A, B, and C, and even a bit of protein?

The route to a better way of living in America is to grow a garden, store the foods, simplify your diet, nibble lower on the food chain—stay close to nature, close to the earth.

APPENDIX A

ORGANIZATIONS AND AGENCIES TO HELP YOU

America is full of experienced gardeners and people with green thumbs. Nearly every community has people who grow beautiful vegetables each year with hardly ever a failure. Often they are willing to share their wisdom and secrets for success. If such a gardener is near and you have specific questions, do not hesitate to seek this kind of local counsel.

Many garden-supply stores are equally friendly and helpful. Since they serve hundreds of gardeners, their information is often the best in the community. Call on them in morning hours when they are not too busy.

U.S. DEPARTMENT OF AGRICULTURE

USDA has been assisting home gardeners for over 100 years. During this time it has produced many useful pamphlets, bulletins, reports, and books on various phases of gardening and home food storage.

The department provides local advice and assistance through approximately 3,100 field offices of the Cooperative Extension Service, usually located in county seats. The chances are good that an Extension Service office exists in your own county; you can usually find it under county listings in your phone book. Even large cities have such offices and services.

These offices have agricultural agents and home economists who are trained in gardening and home food production. Bulletins on gardening, canning, freezing, and storing foods are often available here.

THE NATIONAL GARDEN BUREAU

The National Garden Bureau is a nonprofit organization devoted to the expansion of successful gardening in the United States. It is supported mainly by the seed companies that provide the vegetable and flower seeds sold in garden and hardware stores throughout America. The companies thrive when gardening is successful; therefore they provide useful information through the bureau. The address is:

Derek Fell, Director
National Garden Bureau
Box 1
Gardenville, Pennsylvania 18926

With over twenty years of experience in seeds and gardening in England, Canada, and the United States, Mr. Fell has a good understanding of gardeners' problems in all regions of America. He also serves as director of All-America Selections, the organization that tests and observes varieties of vegetables and flowers, and helps rate them as to usefulness and quality in American gardens. When a seed packet or variety is marked *All-America winner,* this means it merits selection and use by home gardeners.

Derek Fell's views about home gardening are useful:

> My realization that small vegetable gardens can pay big dividends really occurred several years ago when Bill Meachem, editor at that time of *Home Garden* magazine, discussed with me the idea of designing and planting a model vegetable garden. It was grown near Doylestown, Pennsylvania, on a research farm, and we filled it entirely with ten easy-to-grow salad vegetables that could supply an average family over most of the growing season.
>
> It was some surprise to find that the whole garden would fit into an area just 10 feet wide by 15 feet deep. The space was duly staked out, cultivated, limed, fertilized, and planted. The crop of fresh vegetables it produced was astonishing, and

this experiment reinforced my belief in the value of a small vegetable garden.

Until that time I had considered myself a good teacher but a poor gardener. My mistake was in *always* trying to cultivate a garden 50 feet by 50 feet. It was not a question of insufficient enthusiasm—it was simply a matter of not having enough time to keep up with the demands of such a large area.

These past errors qualified me to design a practical small-size vegetable garden. On my own I would not have had the confidence to try it, and the results converted me for life to the wisdom of keeping my vegetable garden small and manageable.

Readers may feel welcome to write to the National Garden Bureau, particularly for information on good varieties of vegetables for their areas.

GARDENS FOR ALL

Gardens For All, Inc., is a nonprofit, tax-exempt educational and counseling organization with a single objective: to promote community gardening.

Based in Vermont, in the past three years Gardens For All has stimulated the formation of community gardens throughout America, including the Burlington, Vermont, community garden program of about 740 plots. They have also served as a national clearing-house for information about community gardens by collecting and sharing the experiences of over 500 community gardens.

Based on "field-tested" successful techniques for starting projects, they have developed three Procedural Manuals that deal with such pragmatic subjects as:

How to line up resource people
How to locate and choose your site
How to get permission to use the land
How to present your case to authority
How to organize
How to register gardeners and assign plots
How to supervise—planting through harvesting
How to finance your project

These manuals are available in three versions: one for charitable institutions, one for schools, and one for profit-making ventures. They cost $10.

Readers may communicate with this organization; they will enthusiastically help you with information you may need. Call Judi Loomis at (802) 863-1371 or write:

Gardens For All, Inc.,
163 Church St.,
Burlington, Vt. 05401

AMERICAN HORTICULTURAL SOCIETY

This fine organization promotes good gardening and care of plants, and accepts memberships from gardeners all over the United States and in other countries. The annual fees are $15. Members receive two bimonthly publications without extra cost: *American Horticulturalist* and *News and Views in Horticulture.*

These magazines arrive in alternate months, and occasionally members also receive samples of new seed varieties.

The American Horticultural Society has recently published the *Directory of American Horticulture,* listing national, state, and local organizations and agencies of interest, with their addresses. This may be obtained for $5.

The address of the Society is:

American Horticultural Society
Mount Vernon,
Virginia 22121

THE AUTHORS ARE GLAD TO ASSIST YOU

Some authors may discourage communication with readers, but we welcome questions, and will be glad to assist you if we feel able to do so.

Please try the above organizations and agencies first, but if not fully served, feel free to write to us at the following address:

Earth Foods Associates
Lee Fryer, President
11221 Markwood Drive
Silver Spring, Maryland 20902

APPENDIX B

WEIGHTS AND MEASURES

Sometimes the directions for applying fertilizer say, "Apply a handful per foot of height of the plant." The problem is that some gardeners have tiny hands that hold about 1 ounce, while others have hands like professional basketball players, which hold half a pound.

The directions may say, "Apply 3 pounds per 100 square feet." But bathroom scales do not weigh 3 pounds, and getting new scales is too expensive.

For these gardeners, the following guides may be useful:

A medium-sized handful of medium-weight commercial fertilizer equals about	2½ ozs.
6 medium-sized handfuls of medium-weight fertilizer equal about	1 lb.
A 2-pound coffee can holds	2 qts. (8 cups)
A 1-pound coffee can holds	1 qt. + 1 oz. (a bit over 4 cups)
1¼ cups limestone or dolomite equals	1 lb.
1 quart dried manure equals	1 lb.
1½ cups rock phosphate equals	1 lb.
3 cups seaweed meal equals	1 lb.
1½ cups bone meal equals	1 lb.
2½ cups cottonseed meal equals	1 lb.
1¼ cups nonorganic garden fertilizer equals	1 lb.

1½ cups dry, organic-based fertilizer equals	1 lb.
2 cups strict organic fertilizer equals	1 lb.
1½ cups Eco-Grow fertilizer equals	1 lb.

A usual metal garden bucket holds 2½ gallons (10 quarts)

28.35 grams equal	1 ounce
453.6 grams equal	1 pound
1 kilogram equals	2.2 pounds
3 teaspoons equal	1 tablespoon
2 tablespoons equal	1 fluid ounce
16 fluid ounces equal	1 pint
32 fluid ounces equal	1 quart

AREAS AND WEIGHTS

43,560 square feet equal 1 acre
A piece of land 208 x 208 equals about 1 acre
1 pound of fertilizer per 100 square feet equals about 440 pounds per acre
5 pounds of fertilizer per 100 square feet equals a bit over 1 ton (2,000 lbs.) per acre

APPENDIX C

A TO Z VEGETABLE GROWING GUIDE

The U.S. Department of Agriculture has published a useful booklet, Bulletin No. 202, "Growing Vegetables in the Home Garden," USDA December 1972. It may be obtained from the U.S. Government Printing Office, Washington, D.C. for seventy-five cents. The following suggestions for growing specific vegetables are taken from this Bulletin, with permission from the Department of Agriculture.

ASPARAGUS

Asparagus is among the earliest of spring vegetables. An area about 20 feet square, or a row of 50 to 75 feet long, will supply plenty of fresh asparagus for a family of five or six persons, provided the soil is well enriched and the plants are given good attention. More must be planted if a supply is to be canned or frozen.

Asparagus does best where winters are cold enough to freeze the ground to a depth of a few inches at last. In many southern areas the plants make a weak growth, producing small shoots. Elevation has some effect, but in general the latitude of south-central Georgia is the southern limit of profitable culture.

The crop can be grown on almost any well-drained, fertile soil, and there is little possibility of having the soil too rich, especially through the use of manure. Loosen the soil far down, either by subsoil plowing or by deep spading before planting. Throw the topsoil aside and spade manure, leaf mold, rotted leaves, or peat into the subsoil to a depth of 14 to 16 inches; then mix from 5 to 10 pounds of a complete fertilizer into each 75-foot row or 20-foot bed.

When the soil is ready for planting, the bottom of the trench should be about 6 inches below the natural level of the soil. After the crowns are set and covered to a depth of an inch or two, gradually work the soil into the trench around the plants during the first season. When set in beds, asparagus plants should be at least 1½ feet apart each way; when set in rows, they should be about 1½ feet apart with the rows from 4 to 5 feet apart.

Asparagus plants, or crowns, are grown from seed. The use of one-year-old plants only is recommended. These should have a root spread of at least 15 inches, and larger ones are better. The home gardener will usually find it best to buy his plants from a grocer who has a good strain of a recognized variety. Mary Washington and Waltham Washington are good varieties that have the added merit of being rust resistant. Waltham Washington is an improved strain of Mary Washington. It contains very little of the purple overcast predominant in the Mary Washington, is a high yielder, and has good green color clear into the ground line. In procuring asparagus crowns it is always well to be sure that they have not been allowed to dry out.

Clean cultivation encourages vigorous growth; it behooves the gardener to keep his asparagus clean from the start. In a large farm garden, with long rows, most of the work can be done with a horse-drawn cultivator or a garden tractor. In a small garden, where the rows are short or the asparagus is planted in beds, however, handwork is necessary.

For a 75-foot row an application of manure and 6 to 8 pounds of a high-grade complete fertilizer, once each year, is recommended. Manure and fertilizer may be applied either before or after the cutting season.

Remove no shoots the year the plants are set in the permanent bed and keep the cutting period short the year after setting. Remove all shoots during the cutting season in subsequent years. Cease cutting about July 1 to 10 and let the tops grow. In the autumn remove and burn the dead tops.

Asparagus rust and asparagus beetles are the chief enemies of the crop.

BEANS

Green beans, both snap and lima, are more important than dry beans to the home gardener. Snap beans cannot be planted until the ground is thoroughly warm, but succession plantings may be made every two weeks from that time until seven or eight weeks before frost. In the lower South and Southwest green beans may be grown during the fall, winter, and spring, but they are not well adapted to midsummer. In the extreme South beans are grown throughout the winter.

Green beans are adapted to a wide range of soils as long as the soils are well drained, reasonably fertile, and of such physical nature that they do not interfere with germination and emergence of the plants. Soil that has received a general application of manure and fertilizer should need no additional fertilization. When beans follow early crops that have been fertilized, the residue of this fertilizer is often sufficient for the beans.

On very heavy lands it is well to cover the planted row with sand, a mixture of sifted coal ashes and sand, peat, leaf mold, or other material that will not bake. Bean seed should be covered not more than 1 inch in heavy soils and 1½ inches in sandy soils. When beans are planted in hills, they may be covered with plant protectors. These covers make it possible to plant somewhat earlier.

Tendercrop, Topcrop, Tenderette, Contender, Harvester, and Kinghorn Wax are good bush varieties of snap beans. Dwarf Horticultural is an outstanding green-shell bean. Brown-seeded or white-seeded Kentucky Wonders are the best pole varieties for snap pods. White Navy, or pea beans, white or red Kidney, and the horticultural types are excellent for dry-shell purposes.

Two types of lima beans, called butter beans in the South, are grown in home gardens. Most of the more northerly parts of the United States, including the northern New England states and the northern parts of other states along the Canadian border, are not adapted to the culture of lima beans. Lima beans need a growing season of about four months with relatively high temperature; they cannot be planted safely until somewhat later than snap beans. The small butter beans mature in a shorter period than the large-seeded lima beans. The use of plant protectors over the seeds is an aid in obtaining earliness.

Lima beans may be grown on almost any fertile, well-drained, mellow soil, but it is especially desirable that the soil be light-textured and not subject to baking, as the seedlings cannot force their way through a hard crust. Covering with some material that will not bake, as suggested for other beans, is a wise precaution when using heavy soils. Lima beans need a soil somewhat richer than is necessary for kidney beans, but the excessive use of fertilizer containing a high percentage of nitrogen should be avoided.

Both the small- and large-seeded lima beans are available in pole and bush varieties. In the South the most commonly grown lima bean varieties are Jackson Wonder, Nemagreen, Henderson Bush, and Sieva pole; in the North, Thorogreen, Dixie Butterpea, and Thaxter are popular small-seeded bush varieties. Fordhook 242 is the most popular midseason large, thick-seeded bush lima bean. King of the Garden and Challenger are the most popular large-seeded pole lima bean varieties.

Pole beans of the kidney and lima types require some form of support, as they normally make vines several feet long. A 5-foot fence makes the best support for pole beans. A more complicated support can be prepared from 8-foot metal fence posts, spaced about 4 feet apart and connected horizontally and diagonally with coarse stout twine to make a trellis. Bean plants usually require some assistance to get started on these supports. Never cultivate or handle bean plants when they are wet; to do so is likely to spread disease.

BEETS

The beet is well adapted to all parts of the country. It is fairly tolerant of heat; it is also resistant to cold. However, it will not withstand severe freezing. In the Northern states, where winters are too severe, the beet is grown in spring, summer, and autumn.

Beets are sensitive to strongly acid soils, and it is wise to apply lime if a test shows the need for it. Good beet quality depends on quick growth; for this the land must be fertile, well-drained, and in good physical condition.

Midsummer heat and drought may interfere with seed germination. By covering the seeds with sandy soil, leaf mold, or other material that will not bake and by keeping the soil damp until the plants are up, much of this trouble can be avoided. Make successive sowings at intervals of about three weeks in order to have a continuous supply of young, tender beets throughout the season.

Where cultivating is by hand, the rows may be about 16 inches apart; where it is by tractor, they must be wider. Beet seed as purchased consists of small balls, each containing several seeds. On most soils the seed should be covered to a depth of about an inch. After the plants are well established, thin them to stand 2 to 3 inches apart in the rows.

Early Wonder, Crosby Egyptian, and Detroit Dark Red are standard varieties suitable for early home-garden planting, while Long Season remains tender and edible over a long season.

BROCCOLI

Heading broccoli is difficult to grow; therefore, only sprouting broccoli is discussed here. Sprouting broccoli forms a loose flower head (on a tall, green, fleshy, branching stalk) instead of a compact head or curd found on cauliflower or heading broccoli. It is one of the newer vegetables in American gardens, but has been grown by Europeans for hundreds of years.

Sprouting broccoli is adapted to winter culture in areas suitable for winter cabbage. It is also tolerant of heat. Spring-set plants in the latitude of Washington, D.C., have yielded good crops of sprouts until midsummer and later under conditions that caused cauliflower to fail. In the latitude of Norfolk, Va., the plant has yielded good crops of sprouts from December until spring.

Sprouting broccoli is grown in the same way as cabbage. Plants grown indoors in the early spring and set in the open about April 1 began to yield sprouts about ten weeks later. The fall crop may be handled in the same way as late cabbage, except that the seed is sown later. The sprouts carrying flower buds are cut about 6 inches long, and other sprouts arise in the axils of the leaves, so that a continuous harvest may be obtained. Green Comet, Calabrese, and Waltham 29 are among the best known varieties.

BRUSSELS SPROUTS

Brussels sprouts are somewhat more hardy than cabbage and will live outdoors over winter in all the milder sections of the country. They may be grown as a winter crop in the South and as early and late as cabbage in the North. The sprouts, or small heads, are formed in the axils (the angle between the leaf stem and the main stalk) of the leaves. As the heads begin to crowd, break the lower leaves from the stem of the plant to give them more room. Always leave the top leaves; the plant needs them to supply nourishment. For winter use in cold areas, take up the plants that are well laden with heads and set them close together in a pit, a cold frame, or a cellar, with some soil tamped around the roots. Keep the stored plants as cool as possible without freezing. Jade Cross, a true F_1 hybrid, has a wide range of adaptability.

CABBAGE

Cabbage ranks as one of the most important home-garden crops. In the lower South it can be grown in all seasons except summer, and in latitudes as far north as Washington, D.C. it is frequently set in the autumn, as its extreme hardiness enables it to live over winter at relatively low temperatures and thus become one of the first spring garden crops. Farther north it can be grown as an early-summer crop and as a late-fall crop for storage. Cabbage can be grown throughout practically the entire United States.

Cabbage is adapted to widely different soils as long as they are fertile, of good texture, and moist. It is a heavy feeder; no vegetable responds better to favorable growing conditions. Quality in cabbage is closely associated with quick growth. Both compost and commercial fertilizer should be liberally used. In addition to the applications made at planting time a side-dressing or two of nitrate of soda, sulfate of ammonia, or other quickly available nitrogenous fertilizer is advisable. These may be applied sparingly to the soil around the plants at intervals of three weeks, not more than 1 pound being used to each 200 square feet of space, or, in terms of single plants, ⅓ ounce to each plant. For late cabbage the supplemental feeding with nitrates may be omitted. Good seed is especially important. Only a few seed is needed for starting enough plants for the home garden, as two or three dozen heads of early cabbage are as many as the average family can use. Early Jersey Wakefield and Golden Acre are standard early sorts. Copenhagen Market and Globe are excellent midseason kinds. Flat Dutch and Danish Ballhead are largely used for late planting.

Where cabbage yellows is a serious disease, resistant varieties should be used. The following are a few of the wilt-resistant varieties adapted to different seasons: Wisconsin Hollander, for late storage; Wisconsin All Seasons, a kraut cabbage, somewhat earlier; Marion Market and Globe, round-head cabbages, for midseason; and Stonehead for an early, small, round-head variety.

Cabbage plants for spring setting in the North may be grown in hotbeds or greenhouses from seeding made a month to six weeks before planting time, or may be purchased from southern growers who produce them outdoors in winter. The winter-grown, hardened plants, sometimes referred to as frostproof, are hardier than hotbed plants and may be set outdoors in most parts of the North as soon as the ground can be worked in the spring. Northern gardeners can have cabbage from their gardens much earlier by using healthy southern-grown plants or well-hardened, well-grown hotbed or greenhouse plants. Late cabbage, prized by northern gardeners for fall use and for storage, is grown from plants produced in open seedbeds from sowings made about a month ahead of planting. Late cabbage may well follow early potatoes, peas, beets, spinach, or other early crop. Many gardeners set cabbage plants between potato rows before the potatoes are ready to dig, thereby gaining time. In protected places, or when plant protectors are used, it is possible always to advance dates somewhat, especially if the plants are well hardened.

CARROTS

Carrots are usually grown in the fall, winter, and spring in the South, providing an almost continuous supply. In the North carrots can be grown and used through the summer and the surplus stored for winter. Carrots will grow on almost any type of soil as long as it is moist, fertile, loose, and free from clods and stones, but sandy loams and peats are best. Use commercial fertilizer.

Because of their hardiness carrots may be seeded as early

in the spring as the ground can be worked. Succession plantings at intervals of three weeks will ensure a continuous supply of tender carrots. Cover carrot seed about ½ inch on most soils; less, usually about ¼ inch, on heavy soils. With care in seeding, little thinning is necessary; carrots can stand some crowding, especially on loose soils. However, they should be no thicker than ten to fifteen plants per foot of row.

Chantenay, Nantes, and Imperator are standard sorts. Carrots should be stored before hard frosts occur, as the roots may be injured by cold.

CAULIFLOWER

Cauliflower is a hardy vegetable, but it will not withstand as much frost as cabbage. Too much warm weather keeps cauliflower from heading. In the South its culture is limited to fall, winter, and spring; in the North, to spring and fall. However, in some areas of high altitude and when conditions are otherwise favorable, cauliflower culture is continuous throughout the summer.

Cauliflower is grown on all types of land from sands to clay and peats. Although the physical character is unimportant, the land must be fertile and well drained. Manure and commercial fertilizer are essential.

The time required for growing cauliflower plants is the same as for cabbage. In the North the main cause of failure with cauliflower in the spring is delay in sowing the seed and setting the plants. The fall crop must be planted at such a time that it will come to the heading stage in cool weather. Snowball and Purple Head are standard varieties of cauliflower. Snow King is an extremely early variety with fair-sized, compact heads of good quality; it has very short stems. Always take care to obtain a good strain of seed; poor cauliflower seed is most objectionable. The Purple Head variety, well adapted for the home garden, turns green when cooked.

A necessary precaution in cauliflower culture with all varieties except Purple Head is to tie the leaves together when the heads, or buttons, begin to form. This keeps the heads white. Cauliflower does not keep long after the heads form; one or two dozen heads are enough for the average garden in one season.

CELERY

Celery can be grown in home gardens in most parts of the country at some time during the year. It is a cool-weather crop and adapted to winter culture in the lower South. In the upper South and in the North it may be grown either as an early-spring or as a late-fall crop. Farther north in certain favored locations it can be grown throughout the summer.

Rich, moist but well-drained, deeply prepared, mellow soil is essential for celery. Soil varying from sand to clay loam and to peat may be used as long as these requirements are met. Unless the ground is very fertile, plenty of organic material, supplemented by liberal applications of commercial fertilizer, is necessary. For a 100-foot row of celery, 5 pounds of a high-grade complete fertilizer thoroughly mixed with the soil are none too much. Prepare the celery row a week or two before setting the plants.

The most common mistake with celery is failure to allow enough time for growing the plants. About ten weeks are needed to grow good celery plants. Celery seed is small and germinates slowly. A good method is to place the seeds in a muslin bag and soak them overnight, then mix them with dry sand, distribute them in shallow trenches in the seed flats or seedbed, and cover them with leaf mold or similar material to a depth of not more than ½ inch. Keep the bed covered with moist burlap sacks. Celery plants are very delicate and must be kept free from weeds. They are made more stocky by being transplanted once before they are set in the garden, but this practice retards their growth. When they are to be transplanted before being set in the ground, the rows in the seed box or seedbed may be only a few inches apart. When they are to remain in the box until transplanted to the garden, however, the plants should be about 2 inches apart each way. In beds the rows should be 10 to 12 inches apart, with seedlings 1 to 1½ inches apart in the row.

For hand culture celery plants are set in rows 18 to 24 inches apart; for tractor cultivation 30 to 36 inches apart. The plants are spaced about 6 inches in the row. Double rows are about a foot apart. Set celery on a cool or cloudy day, if possible; and if the soil is at all dry, water the plants thoroughly. If the plants are large, it is best to pinch off the outer leaves 3 or 4 inches from the base before setting. In bright weather it is well also to shade the plants for a day or two after they are set. Small branches bearing green leaves, stuck in the ground, protect the plants from intense sun without excluding air. As soon as the plants attain some size, gradually work the soil around them to keep them upright. Be careful to get no soil into the hearts of the plants. Early celery is blanched by excluding the light with boards, paper, drain tiles, or other devices. Late celery may be blanched also by banking with earth or by storing in the dark. Banking celery with soil in warm weather causes it to decay.

Late celery may be kept for early-winter use by banking with earth and covering the tops with leaves or straw to keep them from freezing, or it may be dug and stored in a cellar or a cold frame, with the roots well embedded in moist soil. While in storage it must be kept as cool as possible without freezing.

For the home garden Golden Detroit, Summer Pascal (Waltham Improved), and the Golden Plume are adapted for the early crop to be used during late summer, fall, and early winter. For storage and for use after the holiday season, it is desirable to plant some such variety as Green Light or Utah 52–70.

CHARD

Chard, or Swiss chard, is a type of beet that has been developed for its tops instead of its roots. Crop after crop of the outer leaves may be harvested without injuring the plant. Only one planting is necessary, and a row 30 to 40 feet long will supply a family for the entire summer. Each seed cluster contains several seeds, and fairly wide spacing of the seeds facilitates thinning. The culture of chard is practically the same as that of beets, but the plants grow larger and need to be thinned to at least 6 inches apart in the row. Chard needs a rich, mellow soil, and it is sensitive to soil acidity.

CHIVES

Chives are small, onionlike plants that will grow in any place where onions do well. They are frequently planted as a border but are equally well adapted to culture in rows.

Being a perennial, chives should be planted where they can be left for more than one season.

Chives may be started from either seed or clumps of bulbs. Once established, some of the bulbs can be lifted and moved to a new spot. When left in the same place for several years, the plants become too thick; occasionally, dividing and resetting is desirable.

COLLARDS

Collards are grown and used about like cabbage. They withstand heat better than other members of the cabbage group, and are well liked in the South for both summer and winter use. Collards do not form a true head, but a large rosette of leaves, which may be blanched by tying together.

CUCUMBERS

Cucumbers are a warm-weather crop. They may be grown during the warmer months over a wide portion of the country, but are not adapted to winter growing in any but a few of the most southerly locations. Moreover, the extreme heat of midsummer in some places is too severe, and there cucumber culture is limited to spring and autumn.

The cucumber demands an exceedingly fertile, mellow soil high in decomposed organic matter from the compost pile. Also, an additional application of organic matter and commercial fertilizer is advisable under the rows or hills. Be sure the organic matter contains no remains of any vine crops; they might carry injurious diseases. Three or four wheelbarrow loads of well-rotted organic matter and 5 pounds of commercial fertilizer to a 50-foot drill or each ten hills are enough. Mix the organic matter and fertilizer well with the top 8 to 10 inches of soil.

For an early crop, the seed may be started in berry boxes or pots, or on sods in a hotbed, and moved to the garden after danger of late frost is past. During the early growth and in cool periods cucumbers may be covered with plant protectors made of panes of glass with a top of cheesecloth, parchment paper, or muslin. A few hills will supply the needs of a family.

When the seed is planted in drills, the rows should be

6 or 7 feet apart, with the plants thinned to 2 to 3 feet apart in the rows. In the hill method of planting the hills should be a least 6 feet apart each way, with the plants thinned to two in each hill. It is always wise to plant 8 or 10 seeds in each hill, thinned to the desired stand. Cover the seeds to a depth of about ½ inch. If the soil is inclined to bake, cover them with loose earth, such as a mixture of soil and coarse sand, or other material that will not harden and keep the plants from coming through.

When cucumbers are grown primarily for pickling, plant one of the special small-size pickling varieties, such as Chicago Pickling or National Pickling; if they are grown for slicing, plant such varieties as White Spine or Straight Eight. It is usually desirable to plant a few hills of each type; both types can be used for either purpose.

Cucumbers require almost constant vigilance to prevent destructive attacks by cucumber beetles. These insects not only eat the foliage but also spread cucumber wilt and other serious diseases.

Success in growing cucumbers depends largely on the control of diseases and insect pests that attack the crop.

Removal of the fruits before any hard seeds form materially lengthens the life of the plants and increases the size of the crop.

EGGPLANT

Eggplant is extremely sensitive to the conditions under which it is grown. A warm-weather plant, it demands a growing season of from 100 to 140 days with high average day and night temperatures. The soil, also, must be well warmed up before eggplant can safely be set outdoors.

In the South eggplants are grown in spring and autumn; in the North, only in summer. The more northerly areas, where a short growing season and low summer temperatures prevail, are generally unsuitable for eggplants. In very fertile garden soil, which is best for eggplant, a few plants will yield a large number of fruits.

Sow eggplant seeds in a hotbed or greenhouse, or, in warm areas, outdoors about eight weeks before the plants are to be transplanted. It is important that the plants be kept growing without check from low or drying temperatures or other causes. They may be transplanted like tomatoes.

Good plants have stems that are not hard or woody; one with a woody stem rarely develops satisfactorily. Black Beauty, Early Beauty Hybrid, and Jersey King Hybrid are good varieties.

ENDIVE

Endive closely resembles lettuce in its requirements, except that it is less sensitive to heat. It may be substituted for lettuce when the culture of lettuce is impracticable. In the South it is mainly a winter crop. In the North it is grown in spring, summer, and autumn and is also forced in winter. Full Heart Batavian and Salad King are good varieties. Broad-leaved endive is known on the markets as escarole.

Cultural details are the same as those for head lettuce. When the plants are large and well formed, draw the leaves together and tie them so that the heart will blanch. For winter use lift the plants with a ball of earth, place them in a cellar or cold frame where they will not freeze, and tie and blanch them as needed.

ENGLISH PEAS

English peas are a cool-weather crop and should be planted early. In the lower South they are grown at all seasons except summer; farther north, in spring and autumn. In the northern states and at high altitudes, they may be grown from spring until autumn, although in many places summer heat is too severe and the season is practically limited to spring. A few succession plantings may be made at ten-day intervals. The later plantings rarely yield as well as the earlier ones. Planting may be resumed as the cool weather of autumn approaches, but the yield is seldom as satisfactory as that from the spring planting.

Alaska and other smooth-seeded varieties are frequently used for planting in the early spring because of the supposition that they can germinate well in cold, wet soil. Thomas Laxton, Greater Progress, Little Marvel, Freezonia, and Giant Stride are recommended as suitable early varieties with wrinkled seeds. Wando has considerable heat resistance. Alderman and Lincoln are approximately two weeks later than Greater Progress, but under favorable conditions they yield heavily. Alderman is a desirable variety for growing on brush or a trellis. Peas grown on supports are less liable to destruction by birds.

GARLIC

Garlic is more exacting in its cultural requirements than are onions, but it may be grown with a fair degree of success in almost any home garden where good results are obtained with onions.

Garlic is propagated by planting the small cloves, or bulbs, which make up the large bulbs. Each large bulb contains about ten small ones. Carefully separate the small bulbs and plant them singly.

The culture of garlic is practically the same as that of onions. When mature, the bulbs are pulled, dried, and braided into strings or tied in bunches, which are hung in a cool, well-ventilated place.

In the South, where the crop matures early, care must be taken to keep the garlic in a cool, dry place; otherwise it spoils. In the North, where the crop matures later in the season, storage is not so difficult, but care must be taken to prevent freezing.

GOURDS

Gourds have the same general habit of growth as pumpkins and squashes and should have the same general cultural treatment, except that most species require some form of support or trellis to climb upon.

Gourds are used in making dippers, spoons, ladles, salt and sugar containers, and many other kinds of household utensils. They are also used for birdhouses and the manufacture of calabash pipes. But they are of interest chiefly because of their ornamental and decorative possibilities. The thin-shelled, or hard-drying, gourds are the most durable and are the ones that most commonly serve as decorations. The thick-fleshed gourds are more in the nature of pumpkins and squashes, and are almost as perishable.

The thin-shelled gourds of the Lagenaria group are gathered and cured at the time the shells begin to harden, the fruits become lighter in weight, and the tendrils on the vines near the gourds begin to shrivel and dry. For best results give the gourds plenty of time to cure. Some kinds require six months or a year to cure.

The thick-shelled gourds of the Cucurbita group are more difficult to cure than the thin-shelled ones. Their beauty is of short duration; they usually begin to fade after three or four months.

All types of gourds should be handled carefully. Bruises discolor them and cause them to soften and decay.

HORSERADISH

Horseradish is adapted to the north-temperate regions of the United States, but not to the South, except possibly in the high altitudes.

Any good soil, except possibly the lightest sands and heaviest clays, will grow horseradish, but it does best on a deep, rich, moist loam that is well supplied with organic matter. Avoid shallow soil; it produces rough, prongy roots. Mix organic matter with the soil a few months before the plants or cuttings are set. Some fertilizer may be used at the time of planting and more during the subsequent seasons. A top dressing of organic matter each spring is advisable.

Horseradish is propagated either by crowns or by root cuttings. In propagating by crowns a portion of an old plant consisting of a piece of root and crown buds is merely lifted and planted in a new place. Root cuttings are pieces of older roots 6 to 8 inches long and of the thickness of a lead pencil. They may be saved when preparing the larger roots for grating, or they may be purchased from seedsmen. A trench 4 or 5 inches deep is opened with a hoe and the root cuttings are placed at an angle with their tops near the surface of the ground. Plants from these cuttings usually make good roots the first year. As a rule the plants in the home garden are allowed to grow from year to year, and portions of the roots are removed as needed. Pieces of roots and crowns remaining in the soil are usually sufficient to reestablish the plants.

There is very little choice in the matter of varieties of horseradish. Be sure, however, to obtain good healthy planting stock of a strain that is giving good results in the area where it is being grown. New Bohemian is perhaps the best known sort sold by American seedsmen.

KALE

Kale, or borecole, is hardy and lives over winter in latitudes as far north as northern Maryland and southern Pennsylvania and in other areas where similar winter conditions prevail. It is also resistant to heat and may be grown in summer. Its real merit, however, is as a cool-weather green.

Kale is a member of the cabbage family. The best garden varieties are low-growing, spreading plants, with thick, more or less crinkled leaves. Vates Blue Curled, Dwarf Blue Scotch, and Siberian are well-known garden varieties.

No other plant is so well adapted to fall sowing throughout a wide area of both North and South or in areas characterized by winters of moderate severity. Kale may well follow some such early-season vegetable as green beans, potatoes, or peas.

In the autumn the seed may be broadcast very thinly and then lightly raked into the soil. Except for spring sowings, made when weeds are troublesome, sow kale in rows 18 to 24 inches apart and later thin the plants to about a foot apart.

Kale may be harvested either by cutting the entire plant or by taking the larger leaves while young. Old kale is tough and stringy.

KOHLRABI

Kohlrabi is grown for its swollen stem. In the North the early crop may be started like cabbage and transplanted to the garden, but usually it is sown in place. In the South kohlrabi may be grown almost any time except midsummer. The seeds may be started indoors and the plants transplanted in the garden, or the seeds may be drilled in the garden rows and the plants thinned to the desired stand. Kohlrabi has about the same soil and cultural requirements as cabbage, principally a fertile soil and enough moisture. It should be harvested while young and tender. Standard varieties are Purple Vienna and White Vienna.

LEEK

The leek resembles the onion in its adaptability and cultural requirements. Instead of forming a bulb it produces a thick, fleshy cylinder like a large green onion. Leeks are started from seeds, like onions. Usually the seeds are sown in a shallow trench, so that the plants can be more easily hilled up as growth proceeds. Leeks are ready for use any time after they reach the right size. Under favorable conditions they grow to 1½ inches or more in diameter, with white parts 6 to 8 inches long. They may be lifted in the autumn and stored like celery in a cold frame or a cellar.

LETTUCE

Lettuce can be grown in any home garden. It is a cool-weather crop, being as sensitive to heat as any vegetable grown. In the South lettuce culture is confined to late fall, winter, and spring. In colder parts of the South lettuce may not live through the winter. In the North lettuce culture is particularly limited to spring and autumn. In some favored locations, such as areas of high altitude or in far-northern latitudes, lettuce grows to perfection in summer. Planting at the wrong season is responsible for most of the failures with this crop.

Any rich soil is adapted to lettuce, although the plant is sensitive to acid soil. A commercial fertilizer with a heavy proportion of phosphorus is recommended.

Start spring lettuce indoors or in a hotbed and transplant it to the garden when the plants have four or five leaves. Gardeners need not wait for the end of light frost, as lettuce is not usually harmed by a temperature as low as 28°F., if the plants have been properly hardened. Allow about six weeks for growing the plants. For the fall crop the seed may be sown directly in the row and thinned; there is no gain in transplanting.

For tractor cultivation set lettuce plants 12 to 15 inches apart in rows 30 to 36 inches apart; for hand culture, about 14 to 16 inches apart each way. Where gardeners grow leaf lettuce or desire merely the leaves and not well-developed heads, the spacing in the rows may be much closer. In any case it is usually best to cut the entire plant instead of removing the leaves.

There are many excellent varieties of lettuce, all of which do well in the garden when conditions are right. Of the loose-leaf kinds, Black-Seeded Simpson, Grand Rapids, Slobolt, and Saladbowl are among the best. Saladbowl and

Slobolt are heat resistant and very desirable for warm-weather culture. Of the heading sorts, Buttercrunch, White Boston, Fulton, and Great Lakes are among the best. The White Boston requires less time than the three others. Where warm weather comes early, it is seldom worthwhile to sow head lettuce seed in the open ground in the spring with the expectation of obtaining firm heads.

MUSKMELONS

The climatic, soil, and cultural requirements of muskmelons are about the same as for cucumbers, except that they are less tolerant of high humidity and rainy weather. They develop most perfectly on light-textured soils. The plants are vigorous growers and need a somewhat wider spacing than cucumbers.

Hearts of Gold, Hale's Best, and Rocky Ford, the last-named a type, not a variety, are usually grown in the home garden. Where powdery mildew is prevalent, resistant varieties such as Gulf Stream, Dulce, and Perlita are better adapted. Osage and Pride of Wisconsin (Queen of Colorado) are desirable home-garden sorts, particularly in the northern states. Sweet Air (Knight) is a popular sort in the Maryland-Virginia area.

The Casaba and Honey Dew are well adapted only to the West, where they are grown under irrigation.

MUSTARD

Mustard grows well on almost any good soil. As the plants require but a short time to reach the proper stage for use, frequent sowings are recommended. Sow the seeds thickly in drills as early as possible in the spring or, for late use, in September or October. The forms of Indian mustard, the leaves of which are often curled and frilled, are generally used. Southern Curled and Green Wave are common sorts.

OKRA

Okra, or gumbo, has about the same degree of hardiness as cucumbers and tomatoes and may be grown under the same conditions. It thrives on any fertile, well-drained soil. An abundance of quickly available plant food will stimulate growth and ensure a good yield of tender, high-quality pods.

As okra is a warm-weather vegetable, the seeds should not be sown until the soil is warm. The rows should be from 3 to 3½ feet apart, depending on whether the variety is dwarf or large growing. Sow the seeds every few inches and thin the plants to stand 18 inches to 2 feet apart in the rows. Clemson Spineless, Emerald, and Dwarf Green are good varieties. The pods should be picked young and tender, and none allowed to ripen. Old pods are unfit for use and soon exhaust the plant.

ONIONS

Onions thrive under a wide variety of climatic and soil conditions, but do best with an abundance of moisture and a temperate climate, without extremes of heat or cold through the growing season. In the South the onion thrives in the fall, winter, and spring. Farther north winter temperatures may be too severe for certain types. In the North onions are primarily a spring, summer, and fall crop.

Any type of soil will grow onions, but it must be fertile, moist, and in the highest state of tilth. Both compost and commercial fertilizer, especially one high in phosphorus and potash, should be applied to the onion plot. A pound of compost to each square foot of ground and 4 or 5 pounds of fertilizer to each 100 square feet are about right. The soil should be very fine and free from clods and foreign matter.

Onions may be started in the home garden by the use of sets, seedlings, or seed. Sets, or small dry onions grown the previous year—preferably not more than ¾ inch in diameter—are usually employed by home gardeners. Small green plants grown in an outdoor seedbed in the South or in a hotbed or a greenhouse are also in general use. The home-garden culture of onions from seed is satisfactory in the North where the summers are comparatively cool.

Sets and seedlings cost about the same; seeds cost much less. In certainty of results the seedlings are best; practically none form seedstalks. Seed-sown onions are uncertain unless conditions are extremely favorable.

Several distinct types of onions may be grown. The Potato (Multiplier) and Top (Tree) onions are planted in the fall or early spring for use green. Yellow Bermuda, Granex, and White Granex are large, very mild, flat onions for spring harvest in the South; they have short storage life. Sweet Spanish and the hybrids Golden Beauty, Fiesta, Bronze, Perfection, and El Capitan are large, mild, globular onions suited for growing in the middle latitudes of the country; they store moderately well. Southport White Globe, Southport Yellow Globe, Ebenezer, Early Yellow Globe, Yellow Globe Danvers, and the hybrid Abundance are all firm-fleshed long-storage onions for growing as a "main crop" in the Northeast and Midwest. Early Harvest is an early F_1 hybrid adapted to all northern regions of the United States. Varieties that produce bulbs may also be used green.

PARSLEY

Parsley is hardy to cold but sensitive to heat. It thrives under much the same temperature conditions as kale, lettuce, and spinach. If given a little protection, it may be carried over winter through most of the North.

Parsley thrives on any good soil. As the plant is delicate during its early stages of growth, however, the land should be mellow.

Parsley seeds are small and germinate slowly. Soaking in water overnight hastens the germination. In the North it is a good plan to sow the seeds indoors and transplant the plants to the garden, thereby getting a crop before hot weather. In the South it is usually possible to sow the seed directly in drills. For the fall crop in the North row seeding is also practiced. After seeding it is well to lay a board over the row for a few days until the first seedlings appear. After its removal day-to-day watering will ensure germination of as many seeds as possible. Parsley rows should be 14 to 16 inches apart, with the plants 4 to 6 inches apart in the rows. A few feet will supply the family, and a few plants transplanted to the cold frame in the autumn will give a supply during early spring.

PARSNIPS

The parsnip is adapted to culture over a wide portion of the United States. It must have warm soil and weather at planting time, but does not thrive in midsummer in the South.

In many parts of the South parsnips are grown and used during early summer. They should not reach maturity during midsummer, however. Furthermore, it is difficult to obtain good germination in the summer, which limits their culture during the autumn.

Any deep, fertile soil will grow parsnips, but light, friable soil, with no tendency to bake, is best. Stony or lumpy soils are objectionable; they may cause rough, prongy roots.

Parsnip seed must be fresh—not more than a year old–and it is well to sow rather thickly and thin to about 3 inches apart. Parsnips germinate slowly, but it is possible to hasten germination by covering the seed with leaf mold, sand, a mixture of sifted coal ashes and soil, peat, or some similar material that will not bake. Rolling a light soil over the row or trampling it firmly after seeding usually hastens and improves germination. Hollow Crown and All American are suitable varieties.

Parsnips may be dug and stored in a cellar or pit or left in the ground until used. Roots placed in cold storage gain in quality faster than those left in the ground, and freezing in the ground in winter improves the quality.

There is no basis for the belief that parsnips that remain in the ground over winter and start growth in the spring are poisonous. All reported cases of poisoning from eating so-called wild parsnips have been traced to water hemlock (*Cicuta*), which belongs to the same family and resembles the parsnip somewhat.

Be very careful in gathering wild plants that look like the parsnip.

BLACKEYE PEAS

Blackeye peas, also known as cowpeas or southern table peas, are highly nutritious, tasty, and easily grown. Do not plant until danger of frost has passed because they are very susceptible to cold. Leading varieties are Dixilee, Brown Crowder, Lady, Conch, White Acre, Louisiana Purchase, Texas Purple Hull 49, Knuckle Purple Hull, and Monarch Blackeye. Dixilee is a later variety of southern pea. Quality is excellent and it yields considerably more than such old standbys as blackeyes and crowders. It is also quite resistant, or at least tolerant, to nematodes. This fact alone makes it a desirable variety wherever this pest is present. Monarch Blackeye is a fairly new variety of the blackeye type and much better adapted to southern conditions.

Heavy applications of nitrogen fertilizer should not be used for southern peas. Fertilize moderately with a low-nitrogen analysis such as 4–12–12.

For the effort necessary to grow them, few if any other vegetables will pay higher dividends than southern table peas.

PEPPERS

Peppers are more exacting than tomatoes in their requirements, but may be grown over a wide range in the United States. Being hot-weather plants, peppers cannot be planted in the North until the soil has warmed up and all danger of frost is over. In the South planting dates vary with the location, fall planting being practiced in some locations. Start pepper plants six to eight weeks before needed. The seeds and plants require a somewhat higher temperature than those of the tomato. Otherwise they are handled in exactly the same way.

Hot peppers are represented by such varieties as Red Chili and Long Red Cayenne; the mild-flavored by Penn Wonder, Ruby King, World-beater, California Wonder, and Yale Wonder, which mature in the order given.

POTATOES

Potatoes, when grown under favorable conditions, are one of the most productive of all vegetables in terms of food per unit area of land.

Potatoes are a cool-season crop; they do not thrive in midsummer in the southern half of the country. Any mellow, fertile, well-drained soil is suitable for potato production. Stiff, heavy clay soils often produce misshapen tubers. Potatoes respond to a generous use of commercial fertilizer, but if the soil is too heavily limed, the tubers may be scabby.

Commercial 5–8–5 or 5–8–7 mixtures applied at 1,000 to 2,000 pounds to the acre (approximately 7½ to 15 pounds to each 100-foot row) usually provide enough plant food for a heavy crop. The lower rate of application is sufficient for very fertile soils; the higher rate for less fertile ones. Commercial fertilizer can be applied at the time of planting, but it should be mixed with the soil in such a way that the seed pieces will not come in direct contact with it.

In the North plant two types of potatoes—one to provide early potatoes for summer use, the other for storage and winter use. Early varieties include Irish Cobbler, Early Gem, Norland, Norgold, Russet, and Superior. Best late varieties are Katahdin, Kennebec, Chippewa, Russet Burbank, Sebago, and the golden nemotode-resistant Wanseon. Irish Cobbler is the most widely adapted of the early varieties and Katahdin of the late. In the Great Plains States Pontiac and Red La Soda are preferred for summer use, the Katahdin and Russet Burbank for winter. In the Pacific Northwest the Russet Burbank, White Rose, Kennebec, and Early Gem are used. In the southern states the Irish Cobbler, Red La Soda, Red Pontiac, and Pungo are widely grown. The use of certified seed is always advisable.

In preparing seed potatoes for planting, cut them into blocky rather than wedge-shaped pieces. Each piece should be about 1½ ounces in weight and have at least one eye. Medium-sized tubers weighing 5 to 7 ounces are cut to best advantage.

Plant early potatoes as soon as weather and soil conditions permit. Fall preparation of the soil often makes it possible to plant the early crop without delay in late winter or early spring. Potatoes require 2 to 3 weeks to come up, depending on depth of planting and the temperature of the soil. In some sections the ground may freeze slightly, but this is seldom harmful unless the sprouts have emerged. Prolonged cold and wet weather after planting is likely to cause the seed pieces to rot. Hence, avoid too early planting. Young potato plants are often damaged by frost, but they usually renew their growth quickly from uninjured portions of the stems.

Do not dig potatoes intended for storage until the tops are mature. Careful handling to avoid skinning is desirable, and protection from long exposure to light is necessary to prevent their becoming green and unfit for table use. Store in a well-ventilated place where the temperature is low, 45° to 50° if possible, but where there is no danger of freezing.

PUMPKINS

Pumpkins are sensitive to both cold and heat. In the North they cannot be planted until settled weather; in the South they do not thrive during midsummer.

The gardener is seldom justified in devoting any part of a limited garden area to pumpkins, because many other vegetables give greater returns from the same space. However, in gardens where there is plenty of room and where they can follow an early crop like potatoes, pumpkins can often be grown to advantage.

The pumpkin is one of the few vegetables that thrives under partial shade. Therefore it may be grown among sweet corn or other tall plants. Small Sugar and Connecticut Field are well-known orange-yellow-skinned varieties. The Kentucky Field has a grayish-orange rind with salmon flesh. All are good-quality, productive varieties.

Hills of pumpkins, containing one to two plants, should be at least 10 feet apart each way. Pumpkin plants among corn, potato, or other plants usually should be spaced 8 to 10 feet apart in every third or fourth row.

Gather and store pumpkins before they are injured by hard frosts. They keep best in a well-ventilated place where the temperature is a little above 50°F.

RADISHES

Radishes are hardy to cold, but they cannot withstand heat. In the South they do well in autumn, winter, and spring. In the North they may be grown in spring and autumn, and in sections having mild winters they may be grown in cold frames at that season. In high altitudes and in northern locations with cool summers radishes thrive from early spring to late autumn.

Radishes are not sensitive to the type of soil so long as it is rich, moist, and friable. Apply additional fertilizer when the seeds are sown; conditions must be favorable for quick growth. Radishes that grow slowly have a pungent flavor and are undesirable.

Radishes mature the quickest of our garden crops. They remain in prime condition only a few days, which makes small plantings at weekly or ten-day intervals advisable. A few yards of row will supply all the radishes a family will consume during the time the radishes are at their best.

There are two types of radishes—the mild, small, quick-maturing sorts such as Scarlet Globe, French Breakfast, and Cherry Belle, all of which reach edible size in from twenty to forty days, and the more pungent, large, winter radishes such as Long Black Spanish and China Rose, which require seventy-five days or more for growth. Plant winter radishes so they will reach a desirable size in the autumn. Gather and store them like other root crops.

RHUBARB

Rhubarb thrives best in regions having cool, moist summers and winters cold enough to freeze the ground to a depth of several inches. It is not adapted to most parts of the South, but in certain areas of higher elevation it does fairly well. A few hills along the garden fence will supply all that a family can use.

Any deep, well-drained, fertile soil is suitable for rhubarb. Spade the soil or plow it to a depth of 12 to 16 inches and mix in rotted manure, leaf mold, decayed hardwood leaves, sods, or other form of organic matter. The methods of soil preparation suggested for asparagus are suitable for rhubarb. As rhubarb is planted in hills 3 to 4 feet apart, however, it is usually sufficient to prepare each hill separately.

Rhubarb plants may be started from seed and transplanted, but seedlings vary from the parent plant. The usual method of starting the plants is to obtain pieces of crowns from established hills and set them in prepared hills. Top-dress the planting with a heavy application of organic matter in either spring or late fall. Organic matter applied over the hills during early spring greatly hastens growth, or forces the plant.

A pound of complete commercial fertilizer high in nitrogen applied around each hill every year ensures an abundant supply of plant food. The plants can be mulched with green grass or weeds.

Remove seedstalks as soon as they form. No leaf stems should be harvested before the second year and but few until the third. Moreover, the harvest season must be largely confined to early spring. The hills should be divided and reset every seven or eight years; otherwise they become too thick and produce only slender stems.

Crimson, Red Valentine, MacDonald, Canada Red, and Victoria are standard varieties. Use only the leafstalk as a food. *Rhubard leaves contain injurious substances, including oxalic acid. Never use them for food.*

SALSIFY

Salsify, or vegetable oyster, may be grown in practically all parts of the country. It is similar to parsnips in its requirements but needs a slightly longer growing season. For this reason it cannot be grown as far north as parsnips. Salsify, however, is somewhat more hardy and can be sown earlier in the spring.

Thoroughly prepare soil for salsify to a depth of at least a foot. Lighten heavy garden soil by adding sand or comparable material. Salsify must have plenty of plant food.

Sandwich Island is the best-known variety. A half-ounce of seed will sow a 50-foot row, enough for most families. Always use fresh seed; salsify seed retains its vitality only one year.

Salsify may be left in the ground over the winter or lifted and stored like parsnips or other root crops.

SHALLOT

The shallot is a small onion of the Multiplier type. Its bulbs have a more delicate flavor than most onions. Its growth requirements are about the same as those of most other onions. Shallots seldom form seed and are propagated by means of the small cloves, or divisions, into which the plant splits during growth. The plant is hardy and may be left in the ground from year to year, but best results are had by lifting the clusters of bulbs at the end of the growing season and replanting the smaller ones at the desired time.

SPINACH

Spinach is a hardy, cool-weather plant that withstands winter conditions in the South. In most of the North spinach is primarily an early-spring and late-fall crop, but in some areas, where summer temperatures are mild, it may be grown continuously from early spring until late fall. It should be emphasized that summer and winter culture of spinach is possible only where moderate temperatures prevail.

Spinach will grow on almost any well-drained fertile soil where sufficient moisture is available. It is very sensitive to acid soil. If a soil test shows the need, apply lime to the part of the garden used for spinach, regardless of the treatment given the rest of the area.

The application of 100 pounds of rotted manure and 3 to 4 pounds of commercial fertilizer to each 100 square feet of land is suitable for spinach in the home garden. Broadcast both manure and fertilizer and work them in before sowing the seed.

Long Standing Bloomsdale is perhaps the most popular variety seeded in spring. It is attractive, grows quickly, is very productive, and will stand for a moderate length of time before going to seed. Virginia Savoy and Hybrid No. 7 are valuable varieties for fall planting, as they are resistant to yellows, or blight. Hybrid No. 7 is also resistant to downy mildew (blue mold). These two varieties are very cold-hardy but are not suitable for the spring crop, as they produce seedstalks too early. For horse or tractor cultivation the rows of the garden should be not less than 24 inches apart; when land is plentiful, they may be 30 inches apart. For wheel-hoe or handwork the rows should be 14 to 16 inches apart. Spinach may be drilled by hand in furrows about 1 inch deep and covered with fine earth not more than ½ inch deep, or it may be drilled with a seed drill, which distributes the seed more evenly than is ordinarily possible by hand. Thin the plants to 3 or 4 inches apart before they crowd in the row.

SWEET CORN

Sweet corn requires plenty of space and is adapted only to the larger gardens. Although a warm-weather plant, it may be grown in practically all parts of the United States. It needs a fertile, well-drained, moist soil. With these requirements met, the type of the soil does not seem to be especially important, but a clay loam is almost ideal for sweet corn.

In the South sweet corn is planted from early spring

until autumn, but the corn earworm, drought, and heat make it difficult to obtain worthwhile results in midsummer. The ears pass the edible stage very quickly, and succession plantings are necessary to ensure a constant supply. In the North sweet corn cannot be safely planted until the ground has thoroughly warmed up. Here, too, succession plantings need to be made to ensure a steady supply. Sweet corn is frequently planted to good advantage after early potatoes, peas, beets, lettuce, or other early, short-season crops. Sometimes, to gain time, it may be planted before the early crop is removed.

Sweet corn may be grown in either hills or drills, in rows at least 3 feet apart. It is well to plant the seed rather thickly and thin to single stalks 14 to 16 inches apart or three plants to each 3-foot hill. Experiments have shown that in the eastern part of the country there is no advantage in removing suckers from sweet corn. Cultivation sufficient to control weeds is all that is needed.

Hybrid sweet corn varieties, both white and yellow, are usually more productive than the open-pollinated sorts. As a rule they need a more fertile soil and heavier feeding. They should be fertilized with 5–10–5 fertilizer about every three weeks until they start to silk. Many are resistant to disease, particularly bacterial wilt. Never save seed from a hybrid crop for planting. Such seed does not come true to the form of the plants from which it was harvested.

Good yellow-grained hybrids, in the order of the time required to reach edible maturity, are Span-cross, Marcross, Golden Beauty, Golden Cross Bantam, and Ioana. White-grained hybrids are Evergreen and Country Gentleman.

Well-known open-pollinated yellow sorts are Golden Bantam and Golden Midget. Open-pollinated white sorts, in the order of maturity, are Early Evergreen, Country Gentleman, and Stowell Evergreen.

SWEETPOTATO

Sweetpotatoes succeed best in the South, but they are grown in home gardens as far north as southern New York and southern Michigan. They can be grown even farther north, in sections having especially mild climates, such as the Pacific Northwest. In general, sweetpotatoes may be grown wherever there is a frost-free period of about 150 days with relatively high temperature. Jersey Orange, Nugget, and Nemagold are the commonest dry-fleshed varieties; Centennial, Porto Rico, and Goldrush are three of the best of the moist type.

A well-drained, moderately deep sandy loam of medium fertility is best for sweetpotatoes. Heavy clays and very deep, loose-textured soils encourage the formation of long stringy roots. For best results the soil should be moderately fertilized throughout. If applied under the rows, the fertilizer should be well mixed with the soil.

In most of the area over which sweetpotatoes are grown it is necessary to start the plants in a hotbed, because the season is too short to produce a good crop after the weather warms enough to start plants outdoors. Bed roots used for seed close together in a hotbed and cover them with about 2 inches of sand or fine soil, such as leaf mold. It is not safe to set the plants in the open ground until the soil is warm and the weather settled. Toward the last, ventilate the hotbed freely to harden the plants.

The plants are usually set on top of ridges, 3½ to 4 feet apart, with the plants about 12 inches apart in the row. When the vines have covered the ground, no further cultivation is necessary, but some additional hand weeding may be required.

Dig sweetpotatoes a short time before frost, on a bright, drying day when the soil is not too wet to work easily. On a small scale they may be dug with a spading fork, great care being taken not to bruise or injure the roots. Let the roots lie exposed for two or three hours to dry thoroughly; then put them in containers and place them in a warm room to cure. The proper curing temperature is 85°F. Curing for about ten days is followed by storage at 50° to 55°.

SQUASH

Squashes are among the most commonly grown garden plants. They do well in practically all parts of the United States where the soil is fertile and moisture sufficient. Although sensitive to frost, squashes are hardier than melons and cucumbers. In the warmest parts of the South they may be grown in winter. The use of well-rotted composted material thoroughly mixed with the soil is recommended.

There are two classes of squash varieties, summer and

winter. The summer class includes the Bush Scallop, known in some places as the Cymling, the Summer Crookneck, Straightneck, and Zucchini. It also includes the vegetable marrows, of which the best known sort is Italian Vegetable Marrow (Cocozelle). All the summer squashes and the marrows must be used while young and tender, when the rind can be easily penetrated by the thumbnail. The winter squashes include varieties such as Hubbard, Delicious, Table Queen (Acorn), and Boston Marrow. They have hard rinds and are well adapted for storage.

Summer varieties like yellow Straightneck should be gathered before the seeds ripen or the rinds harden, but the winter sorts will not keep unless well matured. They should be taken in before hard frosts and stored in a dry, moderately warm place, such as on shelves in a basement with a furnace. Under favorable conditions such varieties as Hubbard may be kept until midwinter.

TOMATOES

Tomatoes grow under a wide variety of conditions and require only a relatively small space for a large production. Of tropical American origin, the tomato does not thrive in very cool weather. It will, however, grow in winter in home gardens in the extreme South. Over most of the upper South and the North it is suited to spring, summer, and autumn culture. In the more northern areas the growing season is likely to be too short for heavy yields, and it is often desirable to increase earliness and the length of the growing season by starting the plants indoors. By adopting a few precautions, the home gardener can grow tomatoes practically everywhere, given fertile soil with sufficient moisture.

A liberal application of compost and commercial fertilizer in preparing the soil should be sufficient for tomatoes under most conditions. Heavy applications of fertilizer should be broadcast, not applied in the row, but small quantities may be mixed with the soil in the row in preparing for planting.

Start early tomato plants from five to seven weeks before they are to be transplanted to the garden. Enough plants for the home garden may be started in a window box and transplanted to small pots, paper drinking cups with the bottoms removed, plant bands (round or square), or other soil containers. In boxes the seedlings are spaced 2 to 3 inches apart. Tomato seeds germinate best at about 70°F., or ordinary house temperature. Growing tomato seedlings, after the first transplanting, at moderate temperatures, with plenty of ventilation, as in a cold frame, gives stocky, hardy growth. If desired, the plants may be transplanted again to larger containers, such as 4-inch clay pots or quart cans with holes in the bottom.

Tomato plants for all but the early spring crop are usually grown in outdoor seedbeds. Thin seeding and careful weed control will give strong, stocky plants for transplanting.

Tomatoes are sensitive to cold. Never plant them until danger of frost is past. By using plant protectors during cool periods, the home gardener can set tomato plants somewhat earlier than would otherwise be possible. Hot, dry weather, like midsummer in the South, is also unfavorable for planting tomatoes. Planting distances depend on the variety and on whether the plants are to be pruned and staked or not. If pruned to one stem, trained, and tied to stakes or a trellis, they may be set 18 inches apart in 3-foot rows; if not, they may be planted 3 feet apart in rows 4 to 5 feet apart. Pruning and stalking have many advantages for the home gardener. Cultivation is easier, and the fruits are always clean and easy to find. Staked and pruned tomatoes are, however, more subject to losses from blossom-end rot than those allowed to grow naturally.

TURNIPS

Turnips and rutabagas, similar cool-season vegetables, are among the most commonly grown and widely adapted root crops in the United States. They are grown in the South chiefly in the fall, winter, and spring; in the North, largely in the spring and autumn. Rutabagas do best in the more northerly areas; turnips are better for gardens south of the latitude of Indianapolis, Ind., or northern Virginia.

Turnips reach a good size in from sixty to eighty days, but rutabagas need about a month longer. Being susceptible to heat and hardy to cold, these crops should be planted as late as possible for fall use, allowing time for maturity before hard frost. In the South turnips are very popular in the winter and spring. In the North, however, July to August seeding, following early potatoes, peas, or spinach, is the common practice.

Land that has been in a heavily fertilized crop, such as early potatoes, usually gives a good crop without additional fertilizing. The soil need not be prepared deeply, but the surface should be fine and smooth. For spring culture, row planting similar to that described for beets is the best practice. The importance of planting turnips as early as possible for the spring crop is emphasized. When seeding in rows, cover the seeds lightly; when broadcasting, rake the seeds in lightly with a garden rake. A half-ounce of seed will sow a 300-foot row or broadcast 300 square feet. Turnips may be thinned as they grow, and the tops used for greens.

Although there are both white-fleshed and yellow-fleshed varieties of turnips and rutabagas, most turnips are white-fleshed and most rutabagas are yellow-fleshed. Purple Top White Globe and Just Right are the most popular white-fleshed varieties; Golden Ball (Orange Jelly) is the most popular yellow-fleshed variety. American Purple Top is the commonly grown yellow-fleshed rutabaga; Sweet German (White Swede, Sweet Russian) is the most widely used white-fleshed variety. For turnip greens the Seven Top variety is most suitable. This winter-hardy variety overwinters in a majority of locations in the United States.

TURNIP GREENS

Varieties of turnips usually grown for the roots are also planted for the greens. Shogoin is a favorable variety for greens. It is resistant to aphid damage and produces fine-quality white roots if allowed to grow. Seven Top is a leafy sort that produces no edible root. As a rule, sow turnips to be used for greens thickly and then thin them, leaving all but the greens to develop as a root crop. Turnip greens are especially adapted to winter and early-spring culture in the South. The cultural methods employed are the same as those for turnip and rutabaga.

WATERMELONS

Only gardeners with a great deal of space can afford to grow watermelons. Moreover, they are rather particular in their soil requirements, a sand or sandy loam being best. Watermelon hills should be at least 8 feet apart. The plan of mixing a half wheelbarrow load of composted material with the soil in each hill is good, provided the compost is free from the remains of cucurbit plants that might carry diseases. A half-pound of commercial fertilizer also should be thoroughly mixed with the soil in the hill. It is a good plan to place several seeds in a ring about 1 foot in diameter in each hill. Later the plants should be thinned to two to each hill.

New Hampshire Midget, Rhode Island Red, and Charleston Gray are suitable varieties for the home garden. New Hampshire Midget and Sugar Baby are small, extra early, widely grown, very productive varieties. The oval fruits are about 5 inches in diameter; they have crisp, red flesh and dark seeds. Rhode Island Red is an early variety. The fruits are medium in size, striped, and oval; they have a firm rind and bright, pink-red flesh of choice quality. Charleston Gray is a large, long, high-quality, gray-green watermelon with excellent keeping and shipping qualities. It is resistant to anthracnose and fusarium wilt and requires a long growing season.

The preserving type of watermelon—citron—is not edible when raw. Its culture is the same as that for watermelon.

ZUCCHINI

Zucchini squash is a succulent summer vegetable similar to yellow summer squash but of rich green color. It may be boiled and served with butter, salt and pepper, or sliced for addition to green salads.

Plant zucchini when the soil is warm and mellow in the same season as cucumbers, and using the same general and specific directions. Plant the seeds in hills, or use transplants raised in the house or in a cold frame. Dig up the soil and add organic matter and your special fertilizer, under the plants. Then pat it down. Press in the seeds the old-fashioned way with your fingers, ½ to ¾ inches deep, and cover the finger holes. Pat down the soil again. Within ten days, if the weather is warm, the little plants will be up and growing. When they are 2½ to 3 inches high, thin them to 4 plants per hill. When 6 inches high, thin to 2 plants. Zucchini grows 2½ to 3 feet high, and each plant yields many fruit. *Note:*

USDA did not include zucchini in Bulletin No. 202, so we added it. Do not hold them responsible if your zucchinis fail to grow.

INDEX

Index

Index

a guide for examining / ***FRENCH*****
Reading for Meaning

Content *French**: Reading for Meaning* is a supplementary plateau reader suitable for use at the end of the second level of a secondary school French course. It can be used as an adjunct to the A-LM French program or any comparable audio-lingual program. Similar readers are available in German, Russian, and Spanish.

Purpose Native writers have written these selections, carefully controlling grammatical structure and vocabulary so that the student who has completed two levels of classroom instruction can read them easily, with direct association between the printed word and its meaning in French.

Special Features New words are introduced at the rate of about one word in every hundred and are glossed in the right-hand margin. Easily recognizable cognates are marked with a special symbol (*). Footnotes provide appropriate cultural information when necessary. Discussion questions follow each selection, and a complete lexical index follows the main text.

HARCOURT, BRACE & WORLD
New York 10017 / Chicago 60648 / San Francisco 94109
Atlanta 30309 / Dallas 75235